Library of
Davidson College

16 00
5

Sven Danø

Linear Programming in Industry

Theory and Applications

An Introduction

Fourth revised and enlarged edition

Springer-Verlag
New York Wien

Prof. Dr. Sven Danø
Professor of Managerial Economics
University of Copenhagen

This work is subject to copyright.
All rights are reserved, whether the whole or part of the material is concerned, specifically those of translation, reprinting, re-use of illustrations, broadcasting, reproduction by photocopying machine or similar means, and storage in data banks.

© 1974 by Springer-Verlag/Wien
Library of Congress Catalog Card Number 73-13172
Printed in Austria

With 24 Figures

ISBN 0-387-81189-3 Springer-Verlag New York - Wien
ISBN 3-211-81189-3 Springer-Verlag Wien - New York

To my parents

Preface to First Edition

The present volume is intended to serve a twofold purpose. First, it provides a university text of Linear Programming for students of economics or operations research interested in the theory of production and cost and its practical applications; secondly, it is the author's hope that engineers, business executives, managers, and others responsible for the organization and planning of industrial operations may find the book useful as an introduction to Linear Programming methods and techniques. Despite the different backgrounds of these categories of potential readers, their respective fields overlap to a considerable extent; both are concerned with economic optimization problems, and the use of Linear Programming to problems of production planning is simply applied theory of production. The non-economist reader may, but should not, pass over Chapter IV in which the linear production model is linked up with the economic theory of production.

Without being an advanced text, the book aims at covering enough ground to make the reader capable of detecting, formulating, and solving such linear planning problems as he may encounter within his particular field. No heavy demands are made on the reader's mathematical proficiency; except for the proofs in the Appendix—which may be skipped if desired—the mathematical exposition is purely elementary, involving only simple linear relations. In the author's experience, the pedagogical advantages of this approach, as compared with the use of matrix algebra, amply justify the sacrifice of mathematical elegance and typographical simplicity, particularly in explaining the simplex method.

The book is based on lectures given at the University of Copenhagen, the Norwegian College of Agriculture, and the University of Illinois, and on courses held at Dansk Ingeniørforening (The Institution of Danish Civil Engineers). Part of the material presented has been published previously as articles in *Erhvervsøkonomisk Tidsskrift, Metrika, Nordisk Matematisk Tidsskrift, Nordisk Tidsskrift for Industriel Statistik, Nordisk Tidsskrift for Teknisk Økonomi,* and *Operations Research.* I wish to thank the editors for permission to use this material. I am particularly grateful to Professor *David Fog* for valuable editorial criticism, to Professor *David Gale* for the proof of the Fundamental Theorem, and to Mr. *Ernst*

Lykke Jensen, who outlined the inventory model of Chapter V (first published jointly in *Nordisk Tidsskrift for Industriel Statistik*, 1956). I also wish to thank the editors and publishers of *American Economic Review, Econometrica, Harvard Business Review, Journal of Farm Economics*, and *Operations Research*, as well as International Business Machines Corporation, U.S.A., and the Unione Industriale di Torino, for the permission to quote from their publications, and to acknowledge my obligation to the respective authors.

The preparation of the book was made possible by a generous grant from the Graduate Research Board of the University of Illinois, to which my sincere thanks are due.

I am indebted to Miss *Mariam J. Meredith* for her careful and efficient typing of the manuscript.

Copenhagen, April, 1960

Sven Danø

Preface to Fourth Edition

This is a revised and considerably enlarged edition of the book. Two new chapters on integer programming (Chapter VIII) and decomposition (Chapter IX) have been written, and substantial parts of other chapters are new: IV. E on linear investment planning, VI. C on the dual simplex method, and VII. C on parametric programming. Further extensions of the material presented deal with scheduling problems, unbalanced transportation problems, and improvements of the transportation method.

Despite the explosive development in the field of linear programming and mathematical programming in general during the last decade, the author feels more than ever that an introductory text should have an important mission. A basic knowledge of linear programming is a prerequisite of successful application of computers to industrial optimization problems. It is also hoped that the book may serve as an introduction to the vast special literature on linear programming methods.

For suggestions which have led to improvements of the text I am indebted to my students, who have been exposed to previous editions.

Copenhagen, December, 1973

Sven Danø

Contents

I. Introduction 1
   A. Planning Company Operations: The General Problem 1
   B. Linear Planning Models 2
   C. A Simple Example 3

II. Elements of the Mathematical Theory of Linear Programming 5
   A. The Fundamental Theorem 5
   B. The Simplex Method and the Simplex Criterion 12

III. A Practical Example 15

IV. Industrial Applications 27
   A. Blending Problems 27
   B. Optimal Utilization of Machine Capacities 47
   C. Inventory Problems 55
   D. Transportation Problems 60
   E. Linear Investment Planning 66

V. Computational Procedures for Solving Linear Programming Problems ... 75
   A. The Simplex Method 75
   B. The Simplex Tableau 77
   C. Alternate Optima and Second-Best Solutions 84
   D. Computational Short Cuts 87
   E. The Case of Degeneracy 88
   F. Procedure for Solving Transportation Problems 92

VI. Duality in Linear Programming 104
   A. The Duality Theorem 104
   B. Economic Interpretation of the Dual 106
   C. The Dual Simplex Method 109

VII. Sensitivity Analysis and Parametric Programming 112
   A. Sensitivity Analysis 112
   B. A Concrete Example of Sensitivity Analysis 113
   C. Parametric Linear Programming 115

VIII. Integer Linear Programming 121
   A. Integer Programming and Solution by Rounding 121
   B. Solution by Cuts. Pure Case 123
   C. Solution by Cuts. Mixed Case 127
   D. Solution by a Branch-and-Bound Procedure 129

    E. 0−1 Programming 134
    F. Computer Solution 138

IX. Decomposition 139
    A. Decomposition and Decentralized Planning 139
    B. Decomposition by Direct Allocation 144
    C. Decomposition by Shadow Prices 148

X. Appendix 153
    A. Proof of the Fundamental Theorem 153
    B. The Simplex Criterion 154
    C. The Simplex Algorithm 156
    D. Proof of the Duality Theorem 159
    E. Gomory's Algorithms for Integer Programming 162
    F. A Decomposition Theorem 165

References 168

Index 170

I. Introduction

A. Planning Company Operations: The General Problem

At more or less regular intervals, the management of an industrial enterprise is confronted with the problem of *planning operations* for a coming period. Within this category of management problems falls not only the overall planning of the company's aggregate production but problems of a more limited nature such as, for example, figuring the least-cost combination of raw materials for given output or the optimal transportation schedule.

Any such problem of production planning is most rationally solved in two stages:

(i) The first stage is *to determine the feasible alternatives*. For example, what alternative production schedules are at all compatible with the given capacity limitations? What combinations of raw materials satisfy the given quality specifications for the products? etc. The data required for solving this part of the problem are largely of a *technological* nature.

(ii) The second is *to select from among these alternatives one which is economically optimal*: for example, the aggregate production programme which will lead to maximum profit, or the least-cost combination of raw materials. This is where *the economist* comes in; indeed, any economic problem is concerned with making a choice between alternatives, using some criterion of optimal utilization of resources.

Branches of economics such as, e.g., the *theory of production* deal with such problems from a theoretical point of view. In practice, however, problems of operations planning are to a large extent solved by *common-sense methods*, from inspired guesswork to systematic trial-and-error procedures. Any company has its own rules of thumb. It may be difficult to give precise, explicit grounds for their application, let alone "proofs", but there is a general feeling that they lead to at least fairly acceptable results, based as they are upon practical experience, special insight, intuition, and businessman's flair.

It is nevertheless true that a great many problems of operations planning, particularly those of large enterprises, are so complicated that they cannot be solved adequately by common-sense methods. In such cases more *"scientific"* methods come into the picture. The word "scientific" implies in the present context that the concrete problem is formulated in terms of a mathematical model which takes all relevant factors into account and the coefficients of which represent the technological and economic data of the problem; the optimal solution is then computed from the model.

This idea, which is by no means new, is basic to what has since the war become known as *operations research*, a branch of science concerned with the application of mathematical methods to the solution of practical optimization problems in business and other spheres. When used to solve problems of production planning, operations research may be classified as applied theory of production.

B. Linear Planning Models

Linear programming is one of the many different methods that come under the heading of operations research. Mathematically, linear programming is the problem of finding a maximum (or minimum) of a linear function, subject to linear side conditions and to the requirement that the variables should be non-negative. The side conditions form a system of linear equations. When the number of variables exceeds the number of equations the system will in general have an infinite number of solutions, of which those involving negative values of one or more variables are discarded, and the problem is to find the optimal solution, i.e., the one that yields the largest (or smallest) value of the linear function which is used as a criterion of optimality.

It now becomes clear when and how a model of this type is applicable in the theory of production and can be used to solve concrete problems of operations planning. It frequently happens that *the feasible alternatives — the possibilities from which we have to choose, cf. (i) above — appear as alternative solutions to a system of linear equations which are the mathematical expressions of the technological and economic restrictions on the company's freedom of action.* For example, total planned utilization of each machine during the period must be kept within 100% of its capacity; for each product separately, the outputs of the various processes employed shall add up to total stipulated output; and so forth. Moreover, *the economic criterion used by the company for choosing between the alternatives* (cf. (ii) above) *can often be expressed as the maximization* (or minimization) *of a linear function of the same variables.* In many cases maximum profit will be the criterion of optimality and total profit is a linear function of the quantities of inputs and outputs, with their respective prices (assumed to be constant) as coefficients. Similarly, least cost for given output implies the minimization of a linear function of inputs consumed.

The non-negativity requirements, of course, merely serve to guarantee economically meaningful solutions. A solution involving, for example, negative production is obviously inadmissible.

C. A Simple Example

Let us illustrate the linear planning model by a simple concrete example. A brewery has received an order for 100 gallons of 4% beer for prompt delivery. That particular type of beer is not kept in stock so it has to be blended from such types as are available. There are four types in stock and water may be added to dilute the mix if necessary; the respective alcoholic contents of these five possible ingredients and their prices (costs per gallon) are as follows.

	Ingredient no.					Product Specification
	1 (Water)	2	3	4	5	
Alcoholic contents: %	—	2.5	3.7	4.5	5.8	4.0
Price: $/gal.	—	0.22	0.25	0.32	0.45	

The problem is to find a combination of ingredients which will minimize total cost.

Now, since the three first ingredients contain less than 4% of alcohol whereas no. 4 and no. 5 contain more, at least one ingredient from each of these groups must be in the combination. Let us try, for example, to combine no. 1 and no. 4; x_1 and x_4 being the unknown quantities of the two ingredients, this means that we have to solve the equations

$$x_1 + x_4 = 100$$
$$0.045 x_4 = 0.040 (x_1 + x_4) \quad \text{or} \quad 4.5 x_4 = 400$$

where the first equation is the requirement that 100 gallons of mix should be made, whereas the second specifies the alcoholic strength (to which x_1 does not contribute). The solution is

$$x_1 = 11.11, \quad x_4 = 88.89 \text{ (gal.)}$$

and the corresponding cost of ingredients per 100 gallons of mix becomes

$$c = 0.32 \, x_4 = 28.44 \, (\$).$$

the cost of adding water being negligible.

In a similar way we may examine all other combinations of two ingredients, one from each group[1]; the least-cost solution turns out to be

$$x_3 = 62.5, \ x_4 = 37.5; \ c = 27.63.$$

Clearly each of these combinations can be thought of as representing a particular non-negative solution to the system of equations

$$x_1 + \quad x_2 + \quad x_3 + \quad x_4 + \quad x_5 = 100$$
$$0.0\,x_1 + 2.5\,x_2 + 3.7\,x_3 + 4.5\,x_4 + 5.8\,x_5 = 400;$$

blending two ingredients only implies that the other three variables are set $= 0$. For reasons which will become clear later (Ch. II) we need not examine combinations of more ingredients than there are equations. Hence the solution in x_3 and x_4 which we have found is the optimal solution, the one which minimizes total cost of ingredients

$$c = 0.00\,x_1 + 0.22\,x_2 + 0.25\,x_3 + 0.32\,x_4 + 0.45\,x_5.$$

Thus we have solved the problem of finding a non-negative solution to a system of linear equations, which minimizes a linear function of the same variables. This is a typical linear programming problem. In many cases the restrictions will be linear *inequalities* rather than equations, and the problem may be such that the linear function is to be *maximized*. However, as we shall see, all of these variants can be reduced to a common form.

---

[1] Any combination of two ingredients from the *same* group will give a solution which is negative in one of the variables, i.e., non-feasible, as we might expect.

II. Elements of the Mathematical Theory of Linear Programming

A. The Fundamental Theorem

1. The general problem of linear programming can be formulated as follows: Find a set of numbers $x_1, x_2, \ldots, x_n$ which satisfy a system of linear equations (side conditions)

$$a_{11} x_1 + a_{12} x_2 + \ldots + a_{1n} x_n = b_1$$
$$a_{21} x_1 + a_{22} x_2 + \ldots + a_{2n} x_n = b_2 \quad \quad (1\ a)$$
$$\ldots$$
$$a_{m1} x_1 + a_{m2} x_2 + \ldots + a_{mn} x_n = b_m$$

and a set of sign restrictions (non-negativity requirements)

$$x_1 \geq 0, x_2 \geq 0, \ldots, x_n \geq 0 \quad \quad (1\ b)$$

and for which the linear function

$$f = c_1 x_1 + c_2 x_2 + \ldots + c_n x_n \quad \quad (1\ c)$$

has a maximum.

It makes no difference if the linear function is to be *minimized* since this is equivalent to finding a maximum of

$$-f = (-c_1) x_1 + (-c_2) x_2 + \ldots + (-c_n) x_n.$$

In many practical applications the side conditions have the form of linear *inequalities* so that the problem becomes

$$a_{i1} x_1 + a_{i2} x_2 + \ldots + a_{in} x_n \leq b_i \quad (i = 1, 2, \ldots, m)$$
$$x_1 \geq 0, x_2 \geq 0, \ldots, x_n \geq 0 \quad \quad (2)$$
$$c_1 x_1 + c_2 x_2 + \ldots + c_n x_n = f = \text{maximum}.$$

This problem can easily be reduced to a problem involving equations (type (1)) by using an additional unknown x_i' for the difference between the right and left sides of each inequality so that (2) now becomes

$$a_{i1} x_1 + a_{i2} x_2 + \ldots + a_{in} x_n + x_i' = b_i \quad (i = 1, 2, \ldots, m)$$
$$x_1 \geq 0, x_2 \geq 0, \ldots, x_n \geq 0, x_i' \geq 0$$
$$c_1 x_1 + c_2 x_2 + \ldots + c_n x_n + 0 x_1' + \ldots + 0 x_m' = f = \text{maximum}.$$

This is a problem in n "*structural variables*" $x_1, x_2, \ldots, x_n$ and m "*slack variables*" $x'_1, x'_2, \ldots, x'_m$; the latter enter with zero coefficients in the linear function f and they must be non-negative since otherwise the inequalities would be reversed.

Similarly the side conditions of the problem

$$a_{i1} x_1 + a_{i2} x_2 + \ldots + a_{in} x_n \geq b_i \quad (i=1, 2, \ldots, m)$$
$$x_1 \geq 0, \; x_2 \geq 0, \ldots, x_n \geq 0 \tag{3}$$
$$c_1 x_1 + c_2 x_2 + \ldots + c_n x_n = f = \text{minimum}$$

can be transformed into equations by subtracting non-negative slack variables on the left sides:

$$a_{i1} x_1 + a_{i2} x_2 + \ldots + a_{in} x_n - x'_i = b_i \quad (i=1, 2, \ldots, m).$$

In economic applications, the set of coefficients of each variable in the side conditions is usually termed an *activity* (or a *process*) and the variable is called the *level* of the activity. The *structural activities* are those which correspond to the structural variables while the coefficients of x'_i — all zero except for the ith element which is 1 or -1 in (2) and (3) respectively — form a *slack* (or *disposal*) *activity*[1]. The linear form to be maximized or minimized is frequently termed the *preference* (or *objective*) *function*.

2. Evidently problems of this nature are incapable of solution by the traditional methods of finding a maximum of a function subject to side conditions[2]. If a finite maximum of f exists at all this is due to the non-negativity requirements for the variables (including the slack variables) and the solution is not characterized by partial derivatives being equal to zero.

The type of problem may be illustrated by the simplest possible case, that of one linear relation (side condition) in two variables only:

$$a_1 x_1 + a_2 x_2 = b$$
$$x_1 \geq 0, \; x_2 \geq 0 \tag{4}$$
$$c_1 x_1 + c_2 x_2 = f = \text{maximum},$$

where a_1, a_2, b, and at least one of the c's are $\neq 0$. Let us assume for convenience that they are all positive. The problem then is to find among the *feasible solutions* — by which we mean the sets of non-negative numbers (x_1, x_2) which satisfy the equation — one which yields a maximum value for the linear form f. Such a solution is called an *optimal solution* (or a *maximum feasible solution*[3]).

---

[1] In the language of matrix algebra, the coefficients form a *matrix* the *column vectors* of which are the activities. The slack activities are unit vectors with a positive or negative sign in (2) and (3) respectively.

[2] Such as the Lagrange method of undetermined multipliers.

[3] A third name, much used in concrete applications of linear programming, is *an optimal programme*.

One feasible solution is, for example,

$$x_1 = 0, \; x_2 = \frac{b}{a_2}; \tag{5}$$

another is

$$x_1 = \frac{b}{a_1}, \; x_2 = 0. \tag{6}$$

Furthermore, any "convex linear combination" (weighted average) of these two will be a solution, i.e.,

$$x_1 = \frac{\theta b}{a_1}, \; x_2 = \frac{(1-\theta)b}{a_2} \quad (0 < \theta < 1), \tag{7}$$

and this exhausts the possibilities.

The optimal solution to (4) can be determined by exploring the set of feasible solutions starting from say (6). In other words we make x_2 positive at the expense of x_1. Solving the linear equation for x_1 in terms of x_2 we have

$$x_1 = \frac{b}{a_1} - \frac{a_2}{a_1} x_2$$

and substituting in the linear form gives f in terms of x_2:

$$f = \frac{c_1 b}{a_1} + \left(c_2 - \frac{c_1 a_2}{a_1} \right) x_2. \tag{8}$$

For $x_2 = 0$ we have the solution (6) in which we started. Whether a "better" solution can be obtained by making x_2 positive will depend on the sign of the coefficient of x_2 in (8). If it is negative, f is a decreasing function of x_2 and any positive value for x_2 will give a smaller value to f; on the other hand, negative values of x_2 are not admissible, hence f is a maximum for $x_2 = 0$. In other words,

$$x_1 = \frac{b}{a_1}, \; x_2 = 0, \; f = \frac{c_1 b}{a_1}$$

is an optimal solution (in fact, the only one). If the coefficient of x_2 in (8) is $=0$, f will not be affected by an increase in x_2 and any feasible solution (x_1, x_2) will be optimal. Finally, if the coefficient is positive an increase in x_2 will improve the solution. However, x_2 cannot be increased indefinitely because then x_1 would eventually become negative; by the linear equation (or the solution for x_1 in terms of x_2) x_1 becomes zero for $x_2 = b/a_2$ and this is the limit which the non-negativity requirement sets to the growth in x_2. Thus the optimal solution in this case is

$$x_1 = 0, \; x_2 = \frac{b}{a_2}, \; f = \frac{c_2 b}{a_2}.$$

These results are illustrated graphically in Fig. 1. The linear side condition is represented by the straight line going through the points A and B but the non-negativity requirements confine us to points in the non-negative quadrant (including the axes). Thus the set of feasible solutions is represented by the line segment AB whose extreme points correspond to the solutions (5) and (6) above. The geometric picture of the linear preference function is a family of parallel straight lines — one for each value of f — with the slope $-c_1/c_2$. Through every point of AB goes such a line and the one which is farthest north-east represents the maximal value of f. According as the sign of the coefficient in (8) is negative, zero, or positive, the lines will be steeper than AB, parallel to it, or less steep; the optimum will accordingly be the point B, any point on AB, or the point A. Fig. 1 shows the last of the three cases.

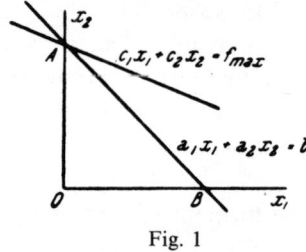

Fig. 1

Had the side condition been an *inequality* of the form

$$a_1 x_1 + a_2 x_2 \leq b, \qquad (9)$$

the set of feasible solutions would have corresponded to the triangle AB 0 in Fig. 1; but we would still be left with the same three possible cases as before. Replacing (9) by

$$a_1 x_1 + a_2 x_2 + x'_1 = b, \qquad (10)$$

where x'_1 is a slack variable, the optimal solution will be

$$x_1 = 0, \ x_2 = \frac{b}{a_2}, \ x'_1 = 0, \ f = \frac{c_2 b}{a_2}$$

when the relative slopes are as shown in the figure. (It takes more than one side condition to get an optimal solution in which a slack variable may be positive.)

Fig. 1 illustrates how the optimal solution appears as a "*corner maximum*"; the geometric picture of the set of feasible solutions is a *convex* area[1] (in this case, the line segment AB or the triangle AB 0) and the

---

[1] Rigorously defined, convexity in a set of points means that the segment joining any two points in the set is also in the set.

optimal solution is one of the *extreme points* ("*corners*") of the area — except for the special case in which any point on the segment AB is optimal, including the two extreme points.

This can be generalized to any number of inequalities in the two structural variables x_1 and x_2. The case of three inequalities is illustrated in Fig. 2 below. The set of feasible solutions is a convex polygon ABCD0 and the optimal point will be one of the corners A, B, C, or D, depending on the ratio of c_1 to c_2. In special cases the solution is not unique and any point on one of the segments AB, BC, or CD respectively is optimal.

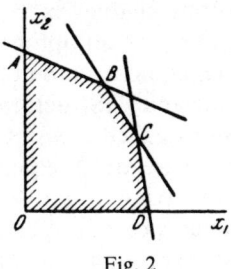

Fig. 2

3. These results provide a clue to the solution of the general linear programming problem.

Fig. 1 is the geometric illustration of a problem in three variables (x_1, x_2, x_1') when the side condition has the form of (10). The area representing the set of feasible solutions has three corners (A, B, and 0) and each corner is characterized by one variable being positive, the remaining two being zero. Any other point on the edge of the area has two positive variables whereas all of the three variables are positive in the interior of the triangle. Similarly Fig. 2 represents a problem in five variables:

$$\begin{aligned} a_{11} x_1 + a_{12} x_2 + x_1' &= b_1 \\ a_{21} x_1 + a_{22} x_2 + x_2' &= b_2 \\ a_{31} x_1 + a_{32} x_2 + x_3' &= b_3. \end{aligned} \quad (11)$$

For $x_1' = 0$, $x_2' = 0$, and $x_3' = 0$ the three equations represent the straight lines containing the segments AB, BC, and CD respectively. In each of the five corners of the polygon ABCD0 three of the variables are positive and the other two equal to zero; for example, the point A is characterized by $x_1 = x_1' = 0$, $x_2 \neq 0$, $x_2' \neq 0$, and $x_3' \neq 0$. Any other point in the convex area has at most one variable equal to zero[1].

Thus in both examples a corner (an extreme point) is characterized by m variables being positive where m is the number of linear side conditions. And we have already seen that the optimal point is one of the corners of the

---

[1] *Exercise:* Show this in detail in Fig. 2.

convex set; even when the optimal solution is not unique it is always possible to find a corner which is an optimal point. Hence in the two cases which we have considered the optimal solution is to be found among those feasible solutions in which precisely m variables assume positive values.

This result can be shown to hold, with a slight qualification, in the general case so that we have the following *Fundamental Theorem of Linear Programming*: *If a linear programming problem with m side conditions has an optimal solution, then there exists such a solution in which at most m of the variables are $\neq 0$.* A general proof of the theorem is given in the Appendix.

4. The Fundamental Theorem enables us to solve the general problem (1), provided that it is known to have an optimal solution. The procedure is to set $n-m$ of the n variables equal to zero in the m linear equations and solve for the remaining m variables. (Structural and slack variables are treated alike.) Such a solution is called a *basic solution*. This can be done in $\binom{n}{n-m} = \binom{n}{m}$ different ways so there are $\binom{n}{m}$ equation systems to be solved; those that yield solutions involving negative values for one or more variables are discarded — in many cases this can be done without actually having to solve them because inspection shows that the solutions will not be feasible — and the optimal solution will be that basic feasible solution which gives the largest value to the preference function.

The optimal solution may contain fewer than m variables >0; such "degenerate" cases will automatically come to light in the form of zero values to some of the m variables in the solution[1]. On the other hand it is perfectly possible that optimal solutions in more than m positive variables exist, but they can be disregarded since the theorem says that no such solution yields a larger value to the preference function than the best basic solution in m variables. In fact such a "non-basic" optimal solution may be expressed as a weighted average of basic optimal solutions in m variables. For example, if (5) and (6) are alternate optimal solutions to (4), (7) too will be optimal.

We will illustrate the procedure by the following numerical example:

$$\begin{array}{l} 2x_1 \quad\quad\quad +4x_3 \geq 5 \\ 2x_1+3x_2+\quad x_3 \geq 4 \end{array} \text{ or } \begin{array}{l} 2x_1 \quad\quad\quad +4x_3-x_1' \quad\quad =5 \\ 2x_1+3x_2+\quad x_3 \quad\quad -x_2'=4 \end{array}$$
$$x_1 \geq 0,\ x_2 \geq 0,\ x_3 \geq 0,\ x_1' \geq 0,\ x_2' \geq 0 \quad\quad (12)$$
$$4x_1+2x_2+3x_3 = g = \text{minimum}.$$

---

[1] For $m=2$ this may happen if b_1 and b_2 are proportional to the coefficients of one of the variables; in the case of two structural variables this means geometrically that the two lines which correspond to the side conditions intersect one of the axes at the same point. For any number of side conditions in two structural variables, a degenerate corner is a point of intersection of more than two of the straight lines which define the set of feasible solutions (including the axes). For example, if points B and C of Fig. 2 happened to coincide, we would have a degenerate feasible solution since at this point only two variables, x_1 and x_2, would be $\neq 0$.

Let us for example set $x_3 = x'_1 = x'_2 = 0$; we call this to choose x_1 and x_2 as *basic variables* (or to choose the corresponding activities as a *basis*). This leaves us with the equation system

$$2 x_1 \qquad = 5$$
$$2 x_1 + 3 x_2 = 4;$$

solving for these variables we get the basic solution

$$x_1 = \frac{5}{2}, \ x_2 = \frac{-1}{3}$$

which is not a feasible solution since x_2 is negative. Next let us try x_1 and x_3 as a basis, i.e., set $x_2 = x'_1 = x'_2 = 0$; the resulting equations

$$2 x_1 + 4 x_3 = 5$$
$$2 x_1 + \ x_3 = 4$$

yield the basic feasible solution

$$x_1 = \frac{11}{6}, \ x_3 = \frac{1}{3}$$

and the corresponding value of the preference function is

$$g = \frac{25}{3}.$$

In this fashion we calculate all of the $\binom{5}{2} = 10$ basic solutions. The results are displayed in the following table.

Basis	Solution	g	Basis	Solution	g
(x_1, x_2)	$(5/2, -1/3)$	—	(x_2, x'_1)	$(4/3, -5)$	—
(x_1, x_3)	$(11/6, 1/3)$	$25/3$	(x_2, x'_2)	no solution	—
(x_2, x_3)	$(11/12, 5/4)$	$67/12$	(x_3, x'_1)	$(4, 11)$	12
(x_1, x'_1)	$(2, -1)$	—	(x_3, x'_2)	$(5/4, -11/4)$	—
(x_1, x'_2)	$(5/2, 1)$	10	(x'_1, x'_2)	$(-5, -4)$	—

The value of g is computed for feasible solutions only. The minimum of g occurs for x_2 and x_3 as basic variables:

$$x_2 = \frac{11}{12}, \ x_3 = \frac{5}{4}, \ g = \frac{67}{12}; \tag{13}$$

provided an optimal solution exists — which we have not yet proved — (13) will be such a solution.

Elements of the Mathematical Theory of Linear Programming

B. The Simplex Method and the Simplex Criterion

For larger values of m and n, however, this procedure soon becomes impracticable because the number of basic solutions to be computed, $\binom{n}{m}$, increases rapidly with m and n. To overcome this difficulty a number of methods have been devised by which the computational labour is reduced considerably. The procedure most frequently used is the so-called *Simplex Method*[1], which is a generalization of the procedure we applied to the trivial problem (4) above. The simplex method is based directly upon the Fundamental Theorem in so far as the procedure is first to choose an arbitrary basic feasible solution as a starting point and next to examine whether a better solution can be obtained by shifting to a neighbouring basis, and so forth, until a basic solution is attained which maximizes (or minimizes) the linear preference function.

To illustrate how the procedure works we shall solve the problem (12) by the simplex method. Let us pick x_1 and x'_2 as an initial basis. Now, instead of setting x_2, x_3, and x'_1 equal to zero as we did above, we treat them as *variables* when solving the equations for x_1 and x'_2:

$$x_1 = \frac{5}{2} \quad -2 x_3 + \frac{1}{2} x'_1$$
$$x'_2 = 1 + 3 x_2 - 3 x_3 + \quad x'_1.$$
(14 a)

Substituting in the linear preference function we get g expressed as a linear function of the non-basic variables:

$$g = 10 + 2 x_2 - 5 x_3 + 2 x'_1.$$
(14 b)

For $x_2 = x_3 = x'_1 = 0$, (14 a—b) gives the positive — and thus feasible — basic solution $x_1 = 5/2$, $x'_2 = 1$, and $g = 10$.

Now whether or not this is an optimal solution will appear from the signs of the coefficients in (14 b), the "*simplex coefficients*". Making x_2 or x'_1 positive will evidently increase the value of the preference function since their coefficients are positive, so the best thing we can do at the present stage is to keep them at zero values. x_3 however, has a negative simplex coefficient so that g is a decreasing function of x_3. This means that a positive value to x_3 will give a better solution. But an increase in x_3 will reduce the values of x_1 and x'_2 and eventually render them negative since, by (14 a), they are both decreasing functions of x_3; x'_2 and x_1 become zero for $x_3 = 1/3$ and $x_3 = 5/4$ respectively (x_2 and x'_1 still $= 0$) and if we increase the value of x_3 beyond these limits this will violate the respective non-negativity requirements. The smaller of the two values is, of course, the

---

[1] The method is due to Dr. George B. Dantzig.

effective limit. At this point we have a new basic feasible solution in which x_3 is $=1/3$ while $x'_2=0$; x_1 is still positive and x_2 and x'_1 are zero as before. Thus x_3 has replaced x'_2 in the basis.

The next step obviously is to solve for the new basic variables x_3 and x_1 and for g in terms of the other three variables. This can be done from the original equations in (12), but it is much easier to find the solution from (14 a). The second equation readily gives us x_3 in terms of the non-basic variables

$$x_3 = \frac{1}{3} + x_2 + \frac{1}{3} x'_1 - \frac{1}{3} x'_2, \tag{15 a}$$

and substitution in the expressions for x_1 and g in (14) yields

$$x_1 = \frac{11}{6} - 2 x_2 - \frac{1}{6} x'_1 + \frac{2}{3} x'_2$$

$$g = \frac{25}{3} - 3 x_2 + \frac{1}{3} x'_1 + \frac{5}{3} x'_2. \tag{15 b}$$

Since x_2 has a negative simplex coefficient in the second equation of (15 b), it is possible to improve the solution further by making x_2 positive, keeping x'_1 and x'_2 at zero. In this case there is only one upper limit to x_2; x_1 becomes zero for $x_2 = 11/12$ whereas x_3 is an increasing function of x_2. Thus the next basis is (x_2, x_3). The first equation of (15 b) gives

$$x_2 = \frac{11}{12} - \frac{1}{2} x_1 - \frac{1}{12} x'_1 + \frac{1}{3} x'_2, \tag{16 a}$$

and substituting in the other two equations of (15) we get

$$x_3 = \frac{5}{4} - \frac{1}{2} x_1 + \frac{1}{4} x'_1$$

$$g = \frac{67}{12} + \frac{3}{2} x_1 + \frac{7}{12} x'_1 + \frac{2}{3} x'_2. \tag{16 b}$$

Since all of the simplex coefficients in the expression for g in (16) are positive no further improvement of the solution is possible and the basic feasible solution

$$x_2 = \frac{11}{12}, x_3 = \frac{5}{4}, g = \frac{67}{12} \tag{17}$$

is optimal.

Thus we have solved the problem in 3 steps, each step corresponding to a basis[1]. Clearly this procedure represents a considerable improvement

---

[1] If we had happened to pick (x_2, x_3) as a starting basis the problem would have been solved in one step.

on the method of computing all of the 10 basic solutions. Once an initial *feasible* solution has been chosen, the logic of the procedure automatically precludes the occurrence of non-feasible basic solutions at the subsequent stages, and the simplex procedure can be thought of as a systematic method of exploring the set of basic feasible solutions without having to compute all of them. Furthermore, as we have seen in the example, it is even possible to get from one solution to the next by a simple substitution; we do not have to solve the original system of equations each time the basis is changed. Moreover, since each step represents a better solution than the previous one, the same basis can never appear more than once in the computations. As the number of bases is finite it follows that the procedure must necessarily lead to an optimal solution in a finite number of steps[1] — provided, of course, that such a solution exists.

In the example above, the optimal solution was characterized by all of the simplex coefficients in (16) being positive. If one or more of them had been zero the solution would still have been optimal (but no longer unique since the value of g is obviously not affected by an increase in a variable whose coefficient in the linear function g is zero). Only when there are negative simplex coefficients will it be possible to improve the solution further. In other words, a *sufficient* condition for a basic feasible solution to be optimal is that all of the simplex coefficients are *non-negative* (in a maximization problem, *non-positive*). This test for optimality is known as the *Simplex Criterion*[2]. It is a *necessary* condition only if the solution is not "degenerate"[3].

The simplex method provides an *iterative* procedure for *numerical* solution of linear programming problems. Explicit *analytical* solution of the general problem (1) is not possible, i.e., the values of the variables x_j and of the preference function f in the optimal solution cannot be expressed as analytical functions of the parameters a_{ij}, b_i, and c_j. This follows from the fact that the optimal solution is a corner solution so that the location of the optimum responds *discontinuously* to changes in the parameters[4].

---

[1] Except for some cases of "degeneracy", cf. Ch. V, E.
[2] See Appendix, B.
[3] See Appendix, B and Ch. V, E.
[4] The effects of parameter changes on the optimal solution are dealt with in some detail in Ch. VII.

III. A Practical Example

1. Linear programming methods have been successfully used for solving many industrial *blending problems*. The simple example in Ch. I belongs to this class of problems and so does the following practical example from the ice cream industry[1].

The first stage in ice cream making is the blending of a number of ingredients (dairy and egg products, sugar, etc.) so as to obtain a mix which meets certain prescribed quality specifications in terms of contents of fat, serum solids, and other constituents.

In general, a mix of desired composition in terms of the constituents can be obtained by many different combinations of ingredients; within this range of substitution, the firm seeks the combination which minimizes total cost of ingredients. As we shall see, the formal structure of this optimization problem is such that linear programming is an appropriate technique for handling it. By this technique, we can simultaneously solve the two problems that are involved: which of the possible ingredients should be used, and in what quantities.

2. The desired composition (in terms of weight) of the mix is as follows:

(1)	Fat	16.00%
(2)	Serum solids	8.00%
(3)	Sugar solids	16.00%
(4)	Egg solids	0.35%
(5)	Stabilizer	0.25%
(6)	Emulsifier	0.15%
(7)	Water	59.25%
	Total	100.00%

The mix is made by blending a number of ingredients each of which contains one or more of these constituents (and nothing else). The following possible ingredients are available on the market, at the prices indicated:

---

[1] The data were kindly provided by Hendrie's Ice Cream Co., Milton, Massachusetts.

A Practical Example

Constituents	Ingredients														Requirements
	1 40% cream	2 23% cream	3 Butter	4 Plastic cream	5 Butter oil	6 4% milk	7 Skim cond. milk	8 Skim milk powder	9 Liquid sugar	10 Sug. froz. egg yolk	11 Powdered egg yolk	12 Stabilizer	13 Emulsifier	14 Water	
(1) Fat	0.400	0.230	0.805	0.800	0.998	0.040				0.500	0.625				16
(2) Serum solids	0.054	0.069		0.025		0.078	0.280	0.970							8
(3) Sugar solids									0.707	0.100					16
(4) Egg solids										0.350	0.315				0.35
(5) Stabilizer												1			0.25
(6) Emulsifier													1		0.15
(7) Total	1	1	1	1	1	1	1	1	1	1	1	1	1	1	100

1. 40% cream .... 0.298 $/lb.
2. 23% cream .... 0.174 $/lb.
3. Butter .... 0.580 $/lb.
4. Plastic cream .... 0.576 $/lb.
5. Butter oil .... 0.718 $/lb.
6. 4% milk .... 0.045 $/lb.
7. Skim condensed milk .... 0.052 $/lb.
8. Skim milk powder .... 0.165 $/lb.
9. Liquid sugar .... 0.061 $/lb.
10. Sugared frozen fresh egg yolk ...... 0.425 $/lb.
11. Powdered egg yolk .... 1.090 $/lb.
12. Stabilizer .... 0.600 $/lb.
13. Emulsifier .... 0.420 $/lb.
14. Water .... —

In the following, the prices of the ingredients, and the quantities in which they are used, are denoted by $c_1, c_2, ..., c_{14}$ and $x_1, x_2, ..., x_{14}$ respectively.

The compositions of these ingredients are shown in the table on p. 16, to be read as follows: 1 unit, say 1 lb., of ingredient no. 1 (40% cream) contains 0.400 lbs. of fat and 0.054 lbs. of serum solids, the residue — not indicated in the table — being water; similarly for the other ingredients. Blank spaces represent zeroes. Stabilizer and emulsifier are constituents as well as ingredients, and therefore, since they contain no water, both have the coefficient 1.

3. We now have the data necessary for figuring the ice cream mix. Taking x_1 lbs. of ingredient no. 1, x_2 lbs. of no. 2, and so forth, to obtain 100 lbs. of mix the composition of which according to the above specifications is indicated in the last column of the table, the following equation must be true for any combination of ingredients (i.e., for any set of x's) that satisfies the fat requirement:

$$0.400\, x_1 + 0.230\, x_2 + 0.805\, x_3 + 0.800\, x_4 \\ + 0.998\, x_5 + 0.040\, x_6 + 0.500\, x_{10} + 0.625\, x_{11} = 16. \qquad (1)$$

Similar equations for the other six requirements, (2)—(7), can be read off the table. The last equation,

$$x_1 + x_2 + ... + x_{14} = 100, \qquad (7)$$

expresses the requirement that the amounts used of the various ingredients add up to 100 lbs. This, of course, is merely another way of stating that the total water content in 100 lbs. of mix must be 59.25 lbs., since — as will be remembered — the water content of each ingredient (and, therefore, of the mix as well) is 100% minus the total percentage of the first six constituents. Clearly (7) is a more convenient form, as it spares us the trouble of computing these residuals[1].

Any combination of ingredients by which a mix of the desired composition can be made is represented by a set of x's satisfying the system of linear equations (1)—(7). Since we have 7 equations in 14 unknowns, i.e., 7 degrees of freedom, the solution is not unique. An example could easily be constructed in which the number of equations and the number of unknowns were equal and the solution was unique, but such a case, leaving no opportunity of choice, would be economically uninteresting; indeed, what makes the present problem an economic one is the multiplicity of solutions. Within this range of substitution, a combination is sought which makes the total cost of ingredients — i.e., the sum of the x's, weighted with their respective prices — as small as possible. Of course, solutions involving one or more negative x's must be disregarded as being economically meaningless, whereas a solution in which there are zeroes implies that some of the ingredients are not used. In other words, the problem is to find a non-negative solution to a system of linear equations, which makes a linear expression in the same variables a minimum.

4. This is exactly the type of problem with which the technique of linear programming is concerned. The table can be considered as a programming matrix; each ingredient defines an activity (the column vector formed by the corresponding set of constant coefficients) and the 14 activities are made interdependent by the requirements expressed in the last column vector.

Formally, the problem may be thought of as a case of joint production, the ingredients being inputs and the constituents being joint outputs. The 14 activities together constitute the production model: when the quantities of the various "outputs" are fixed (specified in the requirements vector), the system represents an isoquant along which a least-cost point is sought[2].

Proceeding to the numerical solving of the problem, inspection of the table shows that we can provisionally discard activities no. 12 and 13, at the same time removing the corresponding equations (5) and (6); stabilizer and emulsifier are "limitational" inputs (having no substitutes within the given technological information as expressed by the matrix), i.e., x_{12} and x_{13} depend on the total amount of ice cream mix to be produced regardless

---

[1] The equation expressing the water content requirement would be
$(1-0.400-0.054) \cdot x_1 + (1-0.230-0.069) \cdot x_2 \ldots + 1 \cdot x_{14} =$
$(100-16-8-16-0.35-0.25-0.15)$ $(=59.25)$. (7a)
Adding equations (1)–(6) to this equation, we obtain (7); equation (7a) then becomes redundant.

[2] Cf. S. Danø (1966), Chs. III and X.

of which combination of the other inputs is chosen. Also, activity no. 14 and equation (7) can for the moment be disregarded since they merely serve to enable us to compute the amount of water which must be added to bring the total weight of the mix up to 100 lbs.

In this manner, the problem has been considerably reduced and we are left with only 4 equations in 11 unknowns. The problem now is:

$$0.400 x_1 + 0.230 x_2 + 0.805 x_3 + 0.800 x_4$$
$$+ 0.998 x_5 + 0.040 x_6 + 0.500 x_{10} + 0.625 x_{11} = 16 \quad (1)$$

$$0.054 x_1 + 0.069 x_2 + 0.025 x_4 + 0.078 x_6$$
$$+ 0.280 x_7 + 0.970 x_8 \qquad\qquad = 8 \quad (2)$$

$$0.707 x_9 + 0.100 x_{10} \qquad\qquad = 16 \quad (3)$$

$$0.350 x_{10} + 0.315 x_{11} \qquad\qquad = 0.35 \quad (4)$$

$$0.298 x_1 + 0.174 x_2 + \ldots + 1.090 x_{11} = c = \text{minimum}$$

where only non-negative values of the x's are admissible.

Now it follows from the Fundamental Theorem that, provided the problem can be solved, there exists a least-cost solution in which no more ingredients are used than there are equations; in other words, we need only examine combinations of four ingredients[1]. There are no less than $\binom{11}{4} = 330$ such combinations, but the simplex technique provides us with a method of finding the best one without actually having to solve all of the corresponding 330 equation systems.

5. The simplex procedure starts with the picking of an initial basis. This can be done arbitrarily, only we must make sure that the corresponding basic solution is positive.

In the present case, it would seem natural to choose as a basis those activities which are actually used by the firm (in this case, ingredients nos. 2, 8, 9, and 10); we may then have a reasonable hope of reaching the optimal combination in a very few steps.

Solving the equations for these basic variables, we get:

$$x_2 = 67.394 - 1.739 x_1 - 3.500 x_3 - 3.478 x_4$$
$$- 4.339 x_5 - 0.174 x_6 - 0.761 x_{11} \qquad (8)$$

$$x_8 = 3.459 + 0.068 x_1 + 0.249 x_3 + 0.221 x_4$$
$$+ 0.308 x_5 - 0.068 x_6 - 0.289 x_7 + 0.054 x_{11} \qquad (9)$$

---

[1] If (1)–(4) had been inequalities, some of the four variables in the optimal basis might have been slack variables, which do not represent ingredients.

$$x_9 = 22.483 + 0.127 \, x_{11} \tag{10}$$

$$x_{10} = 1.000 - 0.900 \, x_{11}. \tag{11}$$

Substituting these expressions in the cost function

$$c = c_1 x_1 + c_2 x_2 + \ldots + c_{11} x_{11},$$

using the ingredient prices given above, we get the total cost of ingredients 1 – 11 as a function of the x's that were not in the basis:

$$\begin{aligned} c = 14.094 \quad &+ 0.007 \, x_1 + 0.012 \, x_3 + 0.007 \, x_4 \\ &+ 0.014 \, x_5 + 0.004 \, x_6 + 0.004 \, x_7 + 0.592 \, x_{11}. \end{aligned}$$

The coefficients in this expression are all positive, i.e., c is an increasing function of the x's. Thus the simplex criterion is satisfied and c assumes its lowest value, 14.094, for all of the x's in the function equal to zero. The corresponding values of the x's in the basis are found by inserting

$$x_1 = x_3 = x_4 = x_5 = x_6 = x_7 = x_{11} = 0$$

in the four equations above; we then get

$$\begin{aligned} x_2 &= 67.394 \\ x_8 &= 3.459 \\ x_9 &= 22.483 \\ x_{10} &= 1.000. \end{aligned}$$

We thus have a non-negative solution which makes the linear expression c a minimum.

In other words, the initial basic solution turned out to be optimal and we have solved the problem already in the first step of the simplex procedure. The solution tells us both which of the ingredients should be used — in this cases nos. 2, 8, 9, and 10 — and in what amounts.

To the total cost of these ingredients, $ 14.094, we must now add the costs of the two limitational inputs, nos. 12 and 13 (whereas there is no additional cost of no. 14, since the price of water is negligible); x_{12} and x_{13} can be read directly off the table (eqs. (5)—(6)) and we thus get:

$$\begin{aligned} c_2 x_2 + c_8 x_8 + c_9 x_9 + c_{10} x_{10} &= 14.094 \\ c_{12} x_{12} = 0.600 \cdot 0.250 \quad &= 0.150 \\ c_{13} x_{13} = 0.420 \cdot 0.150 \quad &= \underline{0.063} \\ & 14.307 \end{aligned}$$

i.e., the total cost of ingredients per 100 lbs. of mix is $ 14.307 if the mix is made in the most economical way (as it actually is).

6. It might be interesting to see how the company has managed to find the least-cost combination by less elaborate methods, and to compare the

linear programming procedure with the methods recommended in the technological literature on ice-cream making.

The simplex technique for computing the least-cost solution has not been used as a practical tool by the company; nor is it mentioned in the technological literature. Nevertheless, both theory and practice have gone a long way in this direction. In a widely used textbook on ice cream making[1], it is implicitly assumed that, in cases where the number of possible inputs exceeds the number of requirements, the least-cost combination of ingredients for making a specified mix is a combination of exactly as many ingredients as there are requirements. In other words, the Fundamental Theorem — a plausible but by no means self-evident proposition — was taken for granted and applied in this particular field before it was actually proved in the development of the linear programming technique. This paved the way for the method of seeking the least-cost combination among the solutions of alternative equation systems each of which has a unique solution; indeed, it is difficult to see by what other method the selection problem could possibly be solved.

The company with which we are concerned does actually figure the mixes by solving simultaneous equations[2]. However, this alone does not carry us very far since we still have not solved the problem of picking, among the great number of possible equation systems (in the present case, 330), the one which represents the optimal combination.

7. Fortunately, in the present case the vast majority of possible combinations can be excluded by systematic inspection of the table. We have already seen how ingredients 12, 13, and 14 can be provisionally disregarded in the solving of the programming problem, thus leaving us with the reduced system (1)−(4). But this is not all. Clearly, we can think of the composition of the mix (the requirements vector) as an *average* of the compositions of the ingredients (the activity vectors). It follows that, in order to get a mix of which 16% is fat, we must use (at least) one input from among those which contain at least 16% fat, i.e., from among the first five ingredients or from nos. 10 and 11; in this sense, ingredients 1—5 and 10—11 are the main sources of fat[3], though, of course, the other ingredients too contribute to the fat content of the mix. Similarly, to satisfy the serum solids requirement (8%) we must use either ingredient 7 or 8, or both. Ingredient no. 9 is the main source of sugar solids, and consequently must be in the optimal combination, since obviously no. 10 does not contain enough sugar solids. Finally, egg solids must be provided by either ingredient 10 or 11, or by both.

---

1 H. H. Sommer (1944), Ch. V: Figuring the Ice Cream Mix.
2 Using a slide rule for the calculations.
3 Cf. H. H. Sommer, op. cit., Ch. V.

By these common-sense considerations, the range of possibilities is narrowed down considerably[1]. The ingredients are separated into groups which are mutually complementary, and the members of each group are mutually substitutable inputs.

It might seem that, since there are four such groups of inputs and four requirements, an optimal solution exists which is a combination of one ingredient from each group and that, consequently, ingredient no. 6, which is not in any of the four groups, can be discarded. However, two of the groups overlap and we must allow for the possibility that, in the optimal combination, two groups are represented by one activity (i.e., by either 10 or 11) while each of the two other groups are represented by one (i.e., 7 or 8, and 9 respectively); in this case activity no. 6 might conceivably be used as a supplementary source of fat and serum solids since, except for the case of degeneracy which is unlikely to occur here, exactly four activities are needed for an optimal combination.

8. Next let us look at the input groups separately. Are there any common-sense methods of figuring out which of the ingredients within a group is to be preferred? We have already seen that, in the *sugar solids* group, ingredient no. 9 will always have to be used. How is the choice made in the other three groups?

In some cases the choice is obvious by inspection. Ingredients 7 and 8 contain nothing but *serum solids* (apart from water, which has no significant economic value and which does not enter into the reduced system (1)–(4)): we can, therefore, convert the prices of the two ingredients into prices of serum solids or — what comes to the same thing — calculate the respective amounts of serum solids which one dollar will purchase, thus computing which ingredient is the cheaper source of this constituent. In the present case, skim milk powder (no. 8) will be preferred, as one dollar spent on this ingredient will purchase 5.9 lbs. of serum solids, as compared with 5.4 lbs. per dollar's worth of skim condensed milk.

In the *egg solids* group (ingredients 10 and 11), such direct comparison is still possible despite the fact that neither ingredient is a pure egg product, since one dollar's worth of sugared egg yolk contains a larger amount of each constituent (viz., 1.2, 0.2, and 0.8 lbs. respectively) than does a dollar's worth of powdered egg yolk (0.6, 0, and 0.3). At the given prices ingredient no. 10 will obviously be preferred as a source of egg solids[2].

Thus we have ascertained that ingredients 8, 9, and 10 must necessarily be in the optimal combination; but we still have not answered the second question: in what quantities? The fourth ingredient will be picked from

---

[1] In the language of linear programming, a great number of non-feasible solutions are excluded, thus narrowing down the range within which a basis is to be sought.

[2] However, no. 11 is still a possible source of fat.

the group of *fat* sources save for the possible case — mentioned above — that ingredient no. 10 (sugared egg yolk) is used as the main source of fat as well as of egg solids, in which case ingredient no. 6 enters into the optimal combination.

Now, in the fat group some of the activities can be discarded by inspection. In the first place, there are two "pure" ingredients: butter (no. 3), and butter oil (no. 5). They happen to be equally economical, since in either case one dollar will purchase 1.39 lbs. of fat; consequently one of them, no matter which, can be excluded from our considerations. Next comparing with plastic cream (no. 4), we see that this latter ingredient gives the same amount of fat per dollar and, in addition, gives an amount of serum solids; it follows that ingredients 3 and 5 will not be used at the given prices.

This means that we are now left with the choice of one from among ingredients 1, 2, 4, 6, and 11, ingredients 8 — 10 being with certainty in the optimal combination and nos. 3, 5, and 7 having been eliminated. No further reduction of the range of choice can be made by the above methods. If, for example, we compare the first two ingredients, we discover that one dollar can purchase less of fat but more of serum solids in ingredient 2 than in no. 1. The same turns out to be true if we compare any other two, as the following table shows:

	Ingredient no.				
	1	2	4	6	11*
Fat, lbs./$	1.34	1.32	1.39	0.89	0.57
Serum solids, lbs./$	0.18	0.40	0.04	1.73	—

* Egg yolk contains egg solids as well.

Evidently this choice problem cannot be solved by simple inspection. Indeed, we are now facing the very type of problem for which the technique of linear programming has been designed.

9. Unless we wish to apply the simplex method, there are two possible methods of dealing with the problem, both of which are mentioned in the technological literature[1] and used in the actual practice of the firm in question. Both methods, incidentally, are implicitly based on linear programming theorems, though they were developed and applied without proof long before the programming technique had been heard of.

The first method is, of course, the solving of each of the equation systems that are obtained by combining ingredients 8 — 10 with ingredients 1, 2, 4, 6, and 11 in turn. Since this gives us only 5 alternative systems to

---
[1] H. H. Sommer, op. cit., Ch. V.

solve, instead of the 330 with which we originally started, it can be done with no great computational labour.

The second method runs as follows: In comparing ingredients 1, 2, 4, and 6, all of which are composed of fat and serum solids, a price is assigned to one of these constituents, e.g., to serum solids; deducting the value of the serum solids content from the market price, we get the "net" cost of fat in each of the ingredients. The problem has thus been reduced to a case of direct comparability (like those above) and that ingredient is chosen which has the lowest net price of fat.

Clearly, the crucial point is what price is to be assigned to the serum solids constituent. Unless the proper price is chosen (provided such a price does exist at all), the procedure is clearly arbitrary and may be misleading. An obvious possibility is to value the serum solids content according to the market price of a pure serum solids product; however, there may not be such a product (though, in the actual case considered, there is); or there may be several products, sold at different prices. Furthermore, an example could easily be constructed in which this rule of thumb would give opposite results according as to whether it is applied to the fat or to the serum solids constituent.

The key to this problem is provided by the so-called Duality Theorem of linear programming. As we shall see later, it follows directly from this theorem that there exists such a set of assigned (imputed) prices which can be computed by solving the "dual" of the programming problem and which, when applied as above, will indicate which ingredients are in the optimal combination and which are not. We shall go further into this in a later chapter (Ch. VI).

10. The procedures which we have described above are based on a number of assumptions some of which might, in certain circumstances, seem somewhat dubious. Let us examine them one by one.

In the first place, it has been tacitly assumed that each constituent is a well-defined substance or, in other words, that the properties of a finished ice cream product (taste, texture, etc.) are the same regardless of which ingredients are used, provided only that the mix satisfies the requirements with respect to gross contents of fat, serum solids, etc. Whether or not this assumption of additivity is justified[1] is of course a purely empirical question (to some extent a matter of taste), which can be settled by practical experience only. It should be noted, however, that regard has been paid to this problem in the selection of possible ingredients. In any case only such ingredients should be considered as can be assumed to make

---

[1] The detailed composition of, e.g., the fat content is not the same in all of the ingredients, but the question in which we are interested is whether such differences affect the taste and the technical properties of the mix.

a mix which is reasonably satisfactory from all relevant technical and qualitative points of view.

11. The next question that arises is whether the composition requirements have to be satisfied exactly, as we have assumed. Would it matter if, for example, the fat content were *larger* than 16 per cent? If the requirements were of the minimal type (inequalities of type $\geq$) instead of being exact equations, it is at least conceivable that a cheaper solution could be found, since the range of feasible solutions would then be considerably larger. It might seem paradoxical that a combination which overfulfils the requirements may be cheaper than an exact solution which involves no "waste" of any constituent; but this is precisely what has actually happened in the solving of an example of the so-called diet problem[1], which is formally identical with the ice cream mix problem with requirements of the minimal type. In such a case, intuition may lead us astray.

Another question is, of course, to what extent the technical properties of the mix and the quality of the finished products are affected by overfulfilment of the contents requirements. According to the experience of the company, the composition of each mix should be kept within rather narrow limits if significant changes in quality — flavour, texture, freezing properties, etc. — are to be avoided. The several types of ice cream which the company produces should be distinguishable, and the quality standards set for each type should be strictly maintained. Nevertheless, it might be worth while to try and determine, by experiment, upper and lower limits between which the composition of the mix could be allowed to vary, and to compute the least-cost solution of this somewhat relaxed programming problem[2].

12. A third problem is that of changing prices. We have assumed prices to be given constants, in spite of the fact that, for example, seasonal fluctuations in prices are met with in practice.

However, such fluctuations are not likely to raise any trouble in the present context, since the production period of ice cream is so short that the input combination can be changed almost immediately, should price changes make it profitable to do so. Indeed, the more frequent prices change, the more useful will our simplex computations turn out to be, as compared with rule-of-thumb methods, since — as we shall see (Ch. VII) — they explicitly allow us to trace the effects of price changes without

---

1 Cf. G. J. Stigler (1945), and T. C. Koopmans (1951), p. 32.

2 We would then have more than four restrictions, at least some of the requirement equations being replaced by two inequalities each (the upper and the lower limit). Since water is added residually the supplementary requirement that the ingredients should add up to 100 lbs. need not be violated.

continually having to recompute the optimal combination from the very beginning by solving further systems of equations.

13. A more serious objection is that, such fluctuations aside, prices cannot always be assumed constant because the price of an ingredient may depend on the quantity to be purchased. An ice cream firm is not likely to be in a monopsonistic position, but price discounts may be granted in case of large purchases. If so, prices can no longer be treated as parameters but must somehow be considered as functions of the x's, and more complicated (nonlinear) methods of solving the least-cost problem must be resorted to. In the present case, however, this complication does not seem to have occurred to any considerable extent; many of the prices are subject to regulations, and the company actually figures the mixes on the basis of fixed input prices.

IV. Industrial Applications

A. Blending Problems

1. One of the first practical problems to be formulated and solved by linear programming methods was the so-called *diet problem*, which is concerned with planning a diet from a given set of foods which will satisfy certain nutritive requirements while keeping the cost at a minimum. For each food the nutritional values in terms of vitamins, calories, etc. per unit of food are known constants and these are the a's of the problem, a_{ij} being the amount of the ith nutritional factor contained in a unit of the jth food. If it is required that there shall be at least b_i units of the ith nutrient in the diet the nutritional requirements will take the form of a set of linear inequalities[1] in the variables x_j, which represent the amounts of the respective foods which shall be present in the diet. These restrictions will in general be satisfied by a large number of combinations of ingredients (foods) and we want to select a combination which minimizes the total cost of ingredients, i.e., a linear function in the x_j where the coefficients c_j are the prices per unit of the respective foods.

The numerical example (12) of Chapter II may be interpreted as a problem of this type. The coefficients of the problem can be arranged in a table as follows.

		"Input" no.			
		1	2	3	
"Output" no.	1	2	0	4	≥ 5
	2	2	3	1	≥ 4
		4	2	3	$= c =$ minimum

From the point of view of production models[2], the three ingredients can be thought of as the inputs of a blending process[3] whose outputs are

---

[1] Nutritional requirements will usually have the form of inequalities since the organism can dispose of excess amounts of the nutrients.

[2] Cf. S. Danø (1966), Chs. III and X.

[3] The term "process" is here used in a broader sense than elsewhere in this book, meaning any transformation of inputs into outputs.

the resulting amounts of the two nutritional factors in the diet. The type of problem is characterized by the coefficient matrix having a column (i.e., an activity) for each input and a row (i.e., a side condition) for each output.

Clearly this type of problem is relevant whenever *dietary problems* are to be solved on a rational basis, taking nutritional and economic factors into account. Solving such problems is part of the daily routine of anyone who runs a large household. However, the same type of problem is frequently met with in *industry*, namely, whenever a product is made by *mechanical blending or mixing of ingredients (raw materials)* and the problem is *to find the cheapest combination of ingredients which satisfies given requirements* with respect to the composition of the mix (in terms of the constituents of which the ingredients are composed).

Our simple hypothetical example in Ch. I belongs to this category except that the specifications were exact (equations). There were two specifications: the alcoholic content of the beer mix should be 4 per cent (or 4 gallons) and the total amount of beer mix should be 100 gallons (which was another way of saying that the total content of all constituents other than alcohol should be 96 per cent or 96 gallons). Thus, by the Fundamental Theorem, all we had to do was to examine all possible combinations of two ingredients.

A concrete industrial example of a similar type, only of larger dimension, is the ice-cream blending problem of Ch. III. The company did not use the simplex method for the practical solution of this problem, but the procedure actually applied anticipated linear programming methods in so far as the optimal solution was found by comparing basic feasible solutions.

Altogether such problems are common in the *food industry*. It is known that, for instance, Armour & Co. has applied linear programming to solve a number of blending problems of the "diet" type and the method appears to be part of the company's general routine. One of these problems — very similar to the ice cream problem — is reported by I. Katzman[1]. It is concerned with finding the least-cost combination of raw materials for making a specified cheese product (processed cheese spread). There are 10 possible ingredients and 7 specifications in the form of inequalities, subject to which the total cost of ingredients is to be minimized. Some of the specifications arise from technical considerations while others represent legal quality requirements. Certain specifications are stated directly in terms of ingredients rather than constituents so the corresponding inequalities have only one coefficient $\neq 0$ (and equal to 1)[2]. Not all of the

---

1 Cf. I. Katzman (1956).
2 Cf. requirements (5) and (6) in the ice cream blending problem.

data are reported but Katzman gives the following example which is part of the complete problem:

Two of the ingredients, sweet cream and curd cheese, contribute to the fat content of the mix, containing 69 and 34% fat respectively. It is a legal requirement that the product shall have a fat content of not less than 20% and that the curd cheese content shall be not less than 51%; furthermore, for technical reasons the maximum content of sweet cream is fixed at 5%. For total output equal to 100 lbs. of the finished product these specifications give rise to the following three restrictions:

$$0.69\,x_1 + 0.34\,x_2 \geq 20$$
$$x_1 \leq 5 \qquad (1)$$
$$x_2 \geq 51$$

where x_1 and x_2 are the quantities of sweet cream and curd cheese in the mix[1].

In addition of these problems of human "diet", linear programming has been applied extensively to solving *feed-mix problems* in agriculture and in the feedstuff industry[2], as well as to calculating the optimal blend of *fertilizers*. These problems, too, are concerned with calculating the least-cost combination of ingredients which satisfies certain nutritional requirements, so they are formally identical with the diet problem.

The manufacturing of *vitamin pills* in the pharmaceutical industry would seem to be another obvious field of application of linear programming. However, vitamin pills are now mostly made from "pure ingredients", i. e., synthetic vitamin preparations, rather than from natural foods which contain more than one vitamin. In this case ingredients and constituents coincide; the problem can still be formulated as a linear programming problem, but it will be a trivial one[3].

It is important to keep in mind that the applicability of linear programming techniques is restricted to problems where the basic assumptions of linearity are really satisfied. This will not be the case where *chemical reactions* occur when the ingredients are mixed; the linear diet problem is concerned with physical (mechanical) blending processes only.

---

[1] *Exercise:* In an (x_1, x_2) diagram, draw the geometric picture of the set of feasible solutions defined by (1). Vary the constants on the right-hand sides in turn to show that not only lower limits ($\geq$) but upper limits as well (inequalities of type $\leq$, such as the second inequality) can be effective restrictions in a minimization problem in the sense that they affect the position of the optimal point. (Consider this as a problem in x_1 and x_2 only, i. e., as a problem of minimizing a linear form in these two variables subject to (1), thus neglecting the other variables and restrictions of the complete problem.)

[2] Many of the best known examples are found in the Journal of Farm Economics.

[3] *Exercise:* Show this by constructing an example.

However, the problem may be such that *only certain combinations of ingredients react chemically*; in this case, although linear programming methods cannot solve the complete problem, they may nevertheless give some guidance as to where the optimal solution is to be found. The same is true of the case where *certain combinations of raw materials are associated with special costs* in addition to the costs of the individual raw materials and the general cost of the blending process[1]. In such cases not only the solution which would have been optimal if the problem had been strictly linear, but the second best solution, the third best, etc. should be computed[2].

A type of problem which somewhat resembles the diet problem although no blending of ingredients is involved, is that of planning operations in an oil refinery where a number of different crude oils are submitted to fractional distillation. The various crudes yield wholly or partially the same oil products but in different proportions (fractions). For specified total amounts to be produced of the respective products we have a problem of selecting raw materials which is analogous with the diet problem[3]. Similar problems occur in steel and aluminium mills[4], the ingredients being metal alloys.

2. Another type of industrial blending problem to which linear programming methods are applicable is the problem of *allocating scarce raw materials between a number of products which are blends or compounds of the same materials but in different proportions*. We shall first illustrate the problem by a simple hypothetical example.

An oil company produces two grades of gasoline by blending three crude products in the proportions 2:1:2 and 1:1:3, these coefficients indicating the amounts of the respective crude gasolines present in a unit of each of the finished products. The three materials are available in stated maximum quantities, say 10, 6, and 15 units per period (e.g., per day), because of capacity limitations in the company's refinery department[5]. Gross profit per unit of each finished product being $ 3 and $ 2 respectively, the problem is to plan production such as to maximize total profit per period, subject to the restrictions that total consumption of each crude

---

1 This case is formally analogous to the occurrence of setup costs in problems of planning production subject to machine capacity limitations.

2 Cf. Ch. V, C.

3 An example of such a problem (only with additional restrictions on the quantities of crudes available and with two alternate processes for one of the crudes) is given in G. H. Symonds (1955), Ch. 3.

4 Cf. N. V. Reinfeld and W. R. Vogel (1958), pp. 122–125.

5 Thus the three crude gasolines are intermediate products rather than primary raw materials.

gasoline cannot exceed the supply available. The unknowns of the problem (the structural variables) are the quantities to be produced of the two finished products. Evidently this is a linear programming problem[1]:

		Output no.		
		1	2	
Input no.	1	2	1	≤ 10
	2	1	1	≤ 6
	3	2	3	≤ 15
		3	2	$=f=$ maximum

This type of problem is symmetrical with the diet problem in the sense that, whereas the latter has a column (i.e., a structural activity) for each input and a row (a side condition) for each "output", the situation has now been reversed: the outputs are the unknowns of the problem and it is the inputs which are specified. Furthermore, the problem is no longer one of finding the least-cost combination of raw materials but a problem of allocating given materials for maximum profit.

In a problem like this where the number of side conditions exceeds the number of structural variables it will in general be impossible to find a solution which just uses up the supply of each raw material. Barring degeneracy, it follows from the Fundamental Theorem that at least one input will not be used up — in this case the third crude gasoline, the optimal solution being $x_1 = 4$, $x_2 = 2$, $x'_3 = 1$[2]. Thus part of the supply of crude gasoline no. 3 will have to be disposed of as a waste product[3], unless (i) the excess supply can be sold on the market — this will be profitable for any positive price — or (ii) additional amounts of the first two intermediate products can be purchased from other refineries. Whether or not this will improve the situation will obviously depend on the prices[4]. Either case represents a deviation from the original problem. In case (i), x'_3 becomes a structural variable, to be interpreted as the quantity of a

---

1 *Exercise:* What statistical information is needed for calculating the coefficients in the linear preference function (i.e., the unit gross profit coefficients 3 and 2)?

2 *Exercise:* Solve the problem geometrically.

3 When the company's ultimate objective is to maximize profits, the avoidance of material waste, idle machine hours, etc. becomes a secondary purpose. It may very well happen that a solution in which some of the slack variables are positive will be optimal even though there exist positive solutions in structural variables only. In such cases intuition and common-sense considerations may lead to a solution which is irrational from the point of view of profit maximization.

4 The highest prices which the company can afford to pay for these additional supplies will be equal to the simplex coefficients of x'_1 and x'_2, a result which follows from the Duality Theorem. (Cf. Ch. VI below.)

marketable product with a positive coefficient in the linear profit function f, though the "blending ratios" remain 0:0:1. (More generally, all of the crude gasolines may be considered as final saleable outputs.) Case (ii) implies that the capacity limits are no longer valid.

Problems of this type are of considerable practical interest, particularly in the chemical industries. It would perhaps be difficult to find practical examples in which the several products are mere physical mixtures of the same raw materials in different proportions, other than cases of the "same" commodity being produced in a number of different grades or qualities each of which is characterized by a particular recipe because the properties of the product depends on the proportions in which the inputs are mixed. The above example is such a case. However, the model applies not only to mechanical blending processes but to cases where the proportions in which the raw materials enter are chemically determined, as in the following concrete example[1].

A chemical plant produces fourteen outputs, $x_1 - x_{14}$:

1. Ammonia, NH_3
2. Nitric acid, HNO_3
3. Ammonium nitrate, NH_4NO_3
4. Urea, $CO(NH_2)_2$
5. Ammonium carbonate, $(NH_4)_2CO_3$
6. Hydrogen chloride, HCl
7. Ammonium chloride, NH_4Cl
8. Polyvinyl chloride, $(CH_2:CHCl)_n$
9. Sodium hypochlorite, $NaClO$
10. Hydrogen, H_2
11. Chlorine (liquid), Cl_2
12. Sodium hydroxide (caustic soda), $NaOH$
13. Carbon dioxide, CO_2
14. Acetylene, $CH\!:\!CH$

from the following six raw materials, $v_1 - v_6$:

(1) Acetylene, $CH\!:\!CH$
(2) Carbon dioxide, CO_2
(3) Hydrogen, H_2
(4) Chlorine (gaseous), Cl_2
(5) Sodium hydroxide (caustic soda), $NaOH$
(6) Nitrogen, N_2

---

[1] Borrowed from A. Bordin (1954), Ch. III: Un esempio di programmazione lineare nell'industria, by A. Bargoni, B. Giardina, and S. Ricossa. To preserve company confidentiality, and partly for didactic reasons, the authors have adjusted some of the technological and economic data used in the following.

all of which are produced by the company itself — v_1, v_2, v_3, and v_6 by cracking of methane (adding oxygen obtained by fractional distillation of liquid air), and v_3, v_4, and v_5 by electrolysis of sodium chloride in aqueous solution. This is done in two supplementary plants whose capacities consequently set limits to the supplies of the six raw materials. Some of the materials may alternatively be sold as such, i.e., they are final outputs as well as raw materials for making other final products; and some of the outputs are intermediate products as well, cf. Fig. 3.

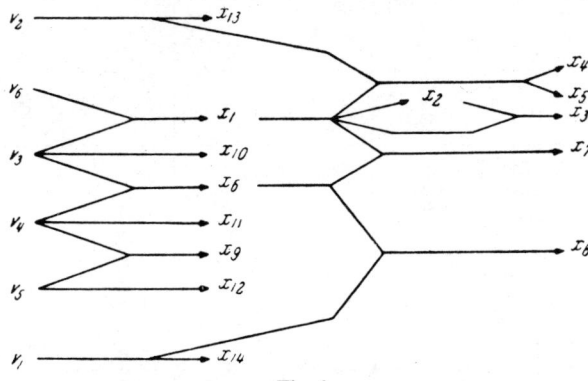

Fig. 3

Since one unit of the final product x_j requires a constant amount of the raw material v_i — directly or by way of intermediate products — the technology of the problem is characterized by a linear model, each final output defining a linear process and the capacity limits for the raw materials giving rise to six linear inequalities. The coefficients of the model are given in the table on p. 34[1]. The problem with which the company is concerned is to determine the quantities to be produced per period (day) of the respective products such as to maximize total gross profit subject to the six side conditions. This is clearly a linear programming problem when gross profit (selling price minus variable costs) per unit of output produced and sold is a constant for each product.

Since five of the raw materials can be sold directly as final products ($x_{10} - x_{14}$) at a positive net profit there is no point in having slack variables associated with them — cf. the gasoline blending problem above — and the corresponding five restrictions have the form of equations in the 14 structural variables.

Further inspection of the table shows that the availability limit for v_6 cannot be an effective bottleneck in the production; v_6 and v_3 are available in the proportion 161,200:130,800 and this ratio is greater than the

---

[1] The units of measurement used are those of the metric system.

	Outputs														Capacity (Supply per day)
	x_1	x_2	x_3	x_4	x_5	x_6	x_7	x_8	x_9	x_{10}	x_{11}	x_{12}	x_{13}	x_{14}	
Units of measurement	t	t	t	t	t	t	t	t	t	m³	kg	kg	kg	kg	
Inputs v_1 kg	2100	320	540	850	510	100	860	440							= 12,800
v_2 kg				1300	780	320	700	210							= 81,600
v_3 m³									180	1.01	1.05				= 130,800
v_4 kg	875	130	235	570	340		290	620	240			1			= 44,300
v_5 kg													1		= 50,000
v_6 m³														1	= 161,200
Gross profit coefficient, lire per unit	90000	15000	9500	60000	30000	8025	85000	200000	17100	5	30	56	40	170	= f = max.
Sales potential, units per day	1	1	∞	3	1	15	1	5	20	∞	3000	∞	3000	8000	

proportion in which the two raw materials enter in any of the products so that v_3 will always be the limiting factor. Consequently we need not pay any attention to the sixth side condition (the sixth row of the table)[1]. This reduction of the problem makes it possible to eliminate some of the products from the picture; v_3 (hydrogen) is now the only scarce raw material used in producing x_1, x_2, x_3, and x_{10}, and if the profit coefficients of these products are divided by their respective hydrogen coefficients we see immediately that x_2 represents the most profitable use of hydrogen[2]. Hence the columns 1, 3, and 10 and the corresponding variables can be disregarded and we are left with a smaller problem involving 5 side conditions in 11 variables only[3,4].

It has so far been assumed that the company can sell all that it can produce. However, in an imperfect market this is not always possible at the given prices. Conditions in the market may be such that the company's sales potential rather than its productive capacity becomes the effective limit to the production programme; if this is evident from the outset the capacity limitations can be ignored and the optimal solution is derived directly from the market restrictions which will have the form of upper limits to the respective x's. Or the sales restrictions may define the set of feasible programme jointly with the capacity limitations[5], in which case the problem will have to be restated with both types of restrictions as side conditions.

In the present example, suppose that the maximum estimated sales potentials for the products are as indicated at the bottom of the table. Only x_3, x_{10}, and x_{12} can be sold in whatever quantities the company can produce. The solution which maximizes total profit subject to the capacity limitations only is not compatible with these additional restrictions so the problem has to be reformulated with the sales limitations as part of the problem. It might be argued that sales could be increased by reducing

---

1 *Exercise:* Explain why an equation whose slack variable is known to be positive can always be disregarded.

2 Cf. the similar procedure by which the ice cream blending problem of Ch. III was partially solved.

3 Using the simplex method, with x_2, x_{11}, x_{12}, x_{13}, and x_{14} as a starting basis, the problem is solved in three steps and the optimal solution turns out to be

$$x_2 = 288.8, \ x_7 = 37.52, \ x_8 = 29.0909, \ x_{12} = 50.0, \ x_{13} = 81.6$$

(all in tons per day), all other variables $= 0$. Total profit is $f = 19{,}403{,}380$ (lire per day).

4 *Exercise:* What price and cost data are needed to compute the coefficients of the profit function? Should the costs of the cracking and electrolysis processes be taken into account?

5 *Exercise:* In the gasoline blending problem above, the introduction of sales restrictions implies the additional side conditions $x_1 \leq \bar{x}_1$ and $x_2 \leq \bar{x}_2$. Show geometrically how this may or may not affect the optimal solution, depending on the values of $\bar{x}_1$ and $\bar{x}_2$.

prices; in the present case, however, the market is such that the prices are fixed by a larger firm which dominates the trade and are taken as data by all other companies, including the one we are considering (price leadership).

Thus we have 11 additional restrictions in the form of inequalities ($x_1 \leq 1$ ton per day, etc.); x_1, x_3, and x_{10} can no longer be disregarded. It takes 11 additional variables to convert them into equations. Furthermore, we no longer assume that the entire supplies of the first five raw materials will be used up or sold, so 5 slack variables are introduced to represent the quantities of $v_1 - v_5$ which are to be disposed of[1]. As a result we now have a problem with 16 side conditions in 30 variables[2], as against 5 side conditions in 11 variables in the previous problem. This is apparently a rather formidable problem, but inspection of the table shows that it can in fact be solved by common-sense methods without using the simplex method.

In the first place, v_1, v_2, and v_4 are no longer effectively scarce factors (bottlenecks). The quantities in which they are available are more than sufficient to produce the maximum quantities that can be sold of the products in which they enter as raw materials; hence the corresponding three capacity restrictions may be disregarded. The two remaining scarce raw materials, v_3 (hydrogen) and v_5 (sodium hydroxide), do not appear together in any column in the table, i.e., in any process. The 9 hydrogen-consuming processes (products) can be arranged in profitability order by calculating profit per m³ of hydrogen used and the solution for the 9 x's follows directly from this order of preference together with the corresponding sales restrictions. The solution for the 2 v_5-consuming products is computed by a similar procedure. Finally, the sales potentials for the remaining 3 products which do not require any scarce materials for their production immediately give the solution for these x's. (In applying the procedure indicated, the profit coefficients of products consuming chlorine (v_4) should be adjusted, taking account of the fact that in producing one unit of the product we save the cost of rendering innocuous the amount of chlorine consumed in making this unit; in other words, the profit coefficient should be increased by $20 \times$ the chlorine coefficient[3].)

---

1 It will serve no useful purpose to store them for later use since they will be in abundant supply also in the following periods so long as the data of the problem remain the same.

These slack variables will have zero coefficients in the profit function, except for chlorine (v_4) which is poisonous and cannot be disposed of costlessly. Excess amounts of chlorine are rendered innocuous by means of calcium hydroxide, i.e., the chlorine is transformed into calcium hypochlorite which is harmless and can be thrown away. The costs of this process give rise to the negative profit coefficient -20.

2 *Exercise:* Write out the new system of equations, or its coefficient matrix.

3 *Exercise:* Solve the problem numerically, using the procedure indicated.

Using this procedure we find an optimal solution which, for those 11 products which are subjected to sales restrictions, precisely corresponds to the respective sales potentials. The quantities to be produced of the other three outputs are computed residually from the capacity restrictions; only x_{10} will be $=0$. Thus we get a positive solution in precisely 13 variables, corresponding to the number of effective restrictions (2 capacity limitations + 11 sales potentials). Total profit will be the sum of the products of the x's and their profit coefficients, *minus* the cost of disposing of the amount of chlorine which is *not* used in carrying out this programme (and which is computed residually from the fourth capacity restriction).

An optimal solution of this type, which involves unutilized amounts of most of the scarce raw materials, evidently calls for the introduction of further products[1] with a view to better utilization of capacity.

3. Still another type of industrial blending problem to which linear programming techniques have been successfully applied is the following which we shall first illustrate by a hypothetical example.

Suppose that two products are made by blending the same three raw materials which are available in limited supplies, say 15, 10, and 55 units per period respectively. No blending proportions are specified for each product (such as in the previous examples), nor are the proportions in which the products are to be made specified in advance (as they were in the diet problem). *Each output can be made from any one of the inputs,* one unit of the product x_j requiring one unit of the raw material v_i, *or by blending them in arbitrary proportions, except that the blending proportions must be such that certain quality specifications for the products are satisfied.* When these specifications can be expressed as linear restrictions on the variables we have a linear optimization model.

The variables of the problem are neither the inputs (v_i) nor the outputs (x_j) but the quantities of each input which enter into the respective outputs, x_{ij}. Then the three "capacity limitations" on the supply of the raw materials can be expressed as the inequalities

$$\begin{aligned} x_{11} + x_{12} &\leq 15 \\ x_{21} + x_{22} &\leq 10 \\ x_{31} + x_{32} &\leq 55. \end{aligned} \qquad (2)$$

Now suppose that the specific gravities of the two products are required to be 0.6 and 0.5 respectively, gravity being an important quality characteristic (as is frequently the case in the chemical industries). The gravity of a mixture will of course depend on the gravities of the ingredients—assumed in this case to be 1.0, 0.8, and 0.4 respectively — and on the proportions

---

[1] Such as, for example, methanol and its derivatives.

in which they are mixed, i. e., on the x_{ij}. By the definition of specific gravity, the dependence will be linear when the x_{ij} are measured in units of volume, provided that no change of total volume occurs when the ingredients are mixed[1]. The gravity of the first product will be

$$\frac{1.0\, x_{11} + 0.8\, x_{21} + 0.4\, x_{31}}{x_{11} + x_{21} + x_{31}},$$

i. e., the weighted arithmetic average of the gravities of the ingredients[2], and if it is required to be 0.6 this gives rise to the linear equation

$$0.4\, x_{11} + 0.2\, x_{21} - 0.2\, x_{31} = 0. \tag{3a}$$

Similarly, for product no. 2 whose gravity is specified to be 0.5 we get

$$0.5\, x_{12} + 0.3\, x_{22} - 0.1\, x_{32} = 0. \tag{3b}$$

Now let profit per unit of product be 4 and 3 ¢ respectively. Then the optimal allocation of the raw materials can be found by maximizing the profit function

$$f = 4\,(x_{11} + x_{21} + x_{31}) + 3\,(x_{12} + x_{22} + x_{32})$$

subject to the restrictions (2)–(3)[3]. The coefficients of this linear programming problem are given in the following table.

		Output no.					
		1			2		
Input no.	1	1			1		≤15
	2		1			1	≤10
	3			1			1	≤55
Quality specification for output no.	1	0.4	0.2	−0.2				=0
	2				0.5	0.3	−0.1	=0
		4	4	4	3	3	3	=f=max.

It is characteristic of this type of blending problem that the blending proportions are not technologically fixed — there is more than one solution to (2)–(3) — but are unknowns to be determined by the optimization procedure.

---

[1] This assumption of additivity — in chemical engineering known as the "Mixture Law", cf. J. H. Perry (1941), p. 616 — is not always satisfied even in cases of physical blending. For example, if alcohol and water are mixed the volume of the mixture will not be the sum of the volumes of the ingredients although there is no chemical reaction.

[2] *Exercise:* Explain why. (Hint: Interpret the numerator and the denominator of the fraction.)

[3] *Exercise:* Would there have been a problem at all if there had been no restrictions other than (2), and if so what would the solution have been?

Problems of this formal type occur in the *petroleum refinery industry* and linear programming methods have proved useful in solving them. Blending of crude oil products into several grades of gasoline each of which must satisfy certain quality specifications can be described in terms of a model analogous with the example above when each of the properties that define quality "blend linearly", i.e., when the quality characteristics of the blend are simple weighted averages of the qualities of the ingredients.

The classic example[1] is the following: A refinery produces three grades of aviation gasoline by blending four components (refinery streams) which are available in limited amounts per day. Amounts of the components not used in making aviation gasoline are used as raw materials for automobile gasoline (blended to specification with other components). There are no fixed recipes (blending ratios) but there are two quality specifications for each product: an upper limit on the admissible *vapour pressure* (RVP, in lbs. per square inch) and a lower limit on the *octane rating* $(1-c$ PN). These two characteristics, which are measures of the volatility and the ignition properties of the gasoline respectively, can be assumed to blend linearly, at least approximately.

The data of the problem are given in the following tables:

Products	Maximum vapour pressure, lbs. per sq. in.	Minimum octane rating	Profit*, $ per barrel
(1) Aviation gasoline A	6.9	80.0	4.908
(2) Aviation gasoline B	6.9	91.0	5.437
(3) Aviation gasoline C	6.9	100.0	6.042
(4) Automobile gasoline	(no specifications)		4.548

* Selling (or accounting) price minus direct cost of tetraethyl lead admixed.

Ingredients	Availability, barrels per day	Vapour pressure, lbs. per sq. in.	Octane rating* A	B, C
1. Alkylate	3,800	5.0	94.0	107.5
2. Catalytic-cracked gasoline	2,652	8.0	83.0	93.0
3. Straight-run gasoline	4,081	4.0	74.0	87.0
4. Isopentane	1,300	20.5	95.0	108.0

* The octane ratings are higher when the ingredients are blended for aviation gasoline grades B and C because of the larger quantity of tetraethyl lead added (4.0 cm^3 per gallon of finished product, as against 0.5 cm^3 per gallon of Grade A aviation gasoline).

---

1 See A. Charnes, W. W. Cooper, and B. Mellon (1952).

	Output no.				Specifications
	1	2	3	4	
Input no.					
1	1	1	1	1	$= 3{,}800$ ⎫
2	1	1	1	1	$= 2{,}652$ ⎬ barrels
3	1	1	1	1	$= 4{,}081$ ⎬ per day
4	1	1	1	1	$= 1{,}300$ ⎭
"Quality balances" for output no.					
1	14 3 −6 15 1.9 −1.1 2.9 −13.6				≥ 0 ≥ 0
2		16.5 2 −4 17 1.9 −1.1 2.9 −13.6			≥ 0 ≥ 0
3			7.5 −7 −13 8 1.9 −1.1 2.9 −13.6		≥ 0 ≥ 0
Profit, $ per barrel	4.908	5.437	6.042	4.548	$= f =$ maximum

The problem is to determine the optimal values of the 16 unknown x_{ij} subject to the 10 restrictions that can be derived from the data, namely, the four "material balances" for the respective ingredients and two "quality balances" for each of the three grades of aviation gasoline. The optimal production programme is that combination for which total profit.

$$f = 4.908\,(x_{11}+x_{21}+x_{31}+x_{41}) + 5.437\,(x_{12}+x_{22}+x_{32}+x_{42})$$
$$+ 6.042\,(x_{13}+x_{23}+x_{33}+x_{43}) + 4.548\,(x_{14}+x_{24}+x_{34}+x_{44})$$

(in $ per day), has a maximum. The coefficients of this linear programming problem are displayed in the table on p. 40[1, 2, 3].

4. In some cases where it is technologically possible to make a number of products from the same raw materials in arbitrary proportions, a uniform quality of each product can be obtained by spending more on processing when raw materials of inferior quality are used[4]. This is the case in the following (hypothetical) example[5]. A manufacturer produces four products, using three different raw materials (or grades of a single raw

	Tons of material required per ton of product no.				Processing cost, $ per ton of product no.				Material price, $ per ton of material		Material availability, tons per period	
									(a)	(b)	(a)	(b)
	1	2	3	4	1	2	3	4				
Material 1	1.20	1.50	1.50	1.80	18	30	57	54	48	72	100	100
no. 2	1.80	2.25	2.25	2.70	30	60	63	81	24	36	150	150
3	2.00	2.50	2.50	3.00	42	69	66	126	18	24	250	400
Selling price, $ per ton	96	150	135	171								
Sales potential, tons per period	200	100	160	50								

(a) normal price,
(b) premium price.

---

[1] The solution turned out to be degenerate.

[2] *Exercise:* Indicate in what way the table will have to be changed if restrictions on product quantities (sales potentials) are introduced.

[3] Concrete examples of related problems in the refinery industry are given in G. H. Symonds, op. cit., especially Chs. 2 and 4.

[4] Alternatively, inferior grades of the products might be sold at lower prices.

[5] The example is borrowed from A. Henderson and R. Schlaifer (1954), pp. 83f.

material) which are perfect substitutes in the sense that each product can be made from any one of the three materials or from an arbitrary combination ("mixture") of them, but processing cost as well as material cost per unit of product depends on what combination is chosen. The materials are available in limited quantities per period at normal prices but additional (though limited) quantities can be purchased at premium prices. For each product there is an upper limit to sales per period. The table on p. 41 gives the data of the problem.

The problem now is to compute the optimal production programme, i.e., to determine how much is to be produced of each product, and from which raw materials, if total profit per period is to be as great as possible. This is evidently a linear programming problem with 24 structural variables x_{ijk} where i ($=1, 2, 3$) is the number of raw material and j ($=1, 2, 3, 4$) is the product number whereas $k=1$ and $k=2$ indicate that the material is purchased at normal and premium price respectively[1].

5. Of course there exists a wide variety of industrial blending problems which combine the features of the different models we have seen above. As a last example we shall demonstrate a concrete blending problem which is an extended version of the brewery example of Ch. I — a problem of the "diet" type — but which also includes quality specifications of the kind met with in the aviation gasoline blending problem above.

A large Danish brewery wishes to produce a beer, which is characterized by an alcoholic content of 3.1% and by certain other quality specifications, by blending four standard beer types available in stock. Water may be added to balance the mix. The problem is to determine in what proportions the five ingredients shall be mixed in order to minimize total cost of ingredients per 100 hectolitres of product ($=2,641.8$ gallons), subject to the given quality specifications.

The blend is to satisfy the following requirements:

(a) The quantity of beer to be made is 100 hl.
(b) The alcoholic content shall be 3.1%.
(c)—(d) The average gravity before fermentation shall be at least 1.034 and must not exceed 1.040.
(e)—(f) The colour of the blend shall be not less than 8 and not more than 10 EBC units.
(g)—(h) The content of hop resin shall be between 20 and 25 mg iso-humulone per litre of blend.

---

[1] *Exercise*: Use the data given above to set up a table showing the coefficients of the problem. How many restrictions are there? Would there have been any problem if unlimited amounts of the raw materials could have been purchased at normal prices?

The ingredients, of which no. 1 is water, are specified as follows:

	Ingredient no.				
	1	2	3	4	5
Alcoholic content, %	0	2.5	3.7	4.5	5.8
Gravity before fermentation	1	1.030	1.043	1.050	1.064
Colour, EBC units	0	11	9	8	7
Hop resin content, mg per litre	0	30	20	28	30

There are many other beer types in stock, but only light beers (nos. 2 − 5) are considered as possible ingredients since the desired product is a light type. Light beer cannot be made by diluting dark beer because they are fundamentally different in colour and taste.

The problem is to minimize total cost of ingredients,

$$c = 0\,x_1 + 44\,x_2 + 50\,x_3 + 64\,x_4 + 90\,x_5,$$

where the coefficients are their respective prices in Danish crowns per hectolitre[1] (the price of water being negligible), subject to the above requirements.

The first requirement (a) can obviously be expressed as the linear equation

$$x_1 + x_2 + x_3 + x_4 + x_5 = 100. \tag{4a}$$

Now the question is whether the quality specifications too give rise to linear restrictions.

The *alcoholic content* is measured as a percentage by weight. Consequently the alcoholic content of the mix will be the arithmetic average of the alcohol percentages of the ingredients when the x's are measured in *weight units*:

$$\frac{0\,x_1 + 2.5\,x_2 + 3.7\,x_3 + 4.5\,x_4 + 5.8\,x_5}{x_1 + x_2 + x_3 + x_4 + x_5};$$

if it is required to be = 3.1 this gives rise to the linear equation

$$0\,x_1 + 2.5\,x_2 + 3.7\,x_3 + 4.5\,x_4 + 5.8\,x_5 = 310, \tag{4b}$$

the sum of the x's being = 100. When the x's are measured in units of *volume*, as they are in the present case, this equation is an approximation of what we wish to specify; in order to get the exact alcoholic content of the mixture, the x's in the fraction above must be multiplied by the respective gravities of the ingredients (in the numerator as well as in the denominator). However, since the gravities do not differ by more than 1%, the approximation is sufficiently good for all practical purposes —

---

[1] The prices have been adjusted for reasons of confidentiality.

the more so because a tolerance of $\pm 0.1\%$ in the alcoholic strength is allowed for in brewery practice.

The *specific gravity* of the wort — i.e., the gravity of the beer before fermentation — is an important quality characteristic, being a measure of the content of extract present in the liquid to be fermented. Thus the weighted average of the gravities of the ingredients before fermentation indicates the heaviness of the composite beer product[1]. (The gravity of water is, of course, equal to 1.) In the present case the average gravity is specified to be between 1.034 and 1.040 and this gives rise to the inequalities

$$1.000\, x_1 + 1.030\, x_2 + 1.043\, x_3 + 1.050\, x_4 + 1.064\, x_5 \geq 103.4$$
$$1.000\, x_1 + 1.030\, x_2 + 1.043\, x_3 + 1.050\, x_4 + 1.064\, x_5 \leq 104.0.$$

After transforming them into equations by introducing slack variables x_6 and x_7, we can replace the first equation by the difference between the second and the first equation so that we get

$$x_6 + x_7 = 0.6 \quad (4c)$$
$$1.000\, x_1 + 1.030\, x_2 + 1.043\, x_3 + 1.050\, x_4 + 1.064\, x_5 + x_7 = 104.0; \quad (4d)$$

these equations are equivalent to the inequalities above and they are easier to handle since most of the coefficients in the first equation are now zero[2].

The *colour* of a beer is measured, for example, in EBC units, using a photometer. Colour can be assumed to blend linearly so the lower and upper limits can be expressed as linear inequalities. Introducing slack variables x_8 and x_9 and replacing the first restriction by the difference between the two equations (as we did above) we have

$$x_8 + x_9 = 200 \quad (4e)$$
$$0\, x_1 + 11\, x_2 + 9\, x_3 + 8\, x_4 + 7\, x_5 + x_9 = 1{,}000. \quad (4f)$$

Specifying the content of *hop resin* in the mix — as measured in milligrams of isohumulone per litre, or decigrams per hectolitre — is one way of specifying the taste of the product, since the characteristic bitter taste

---

[1] Specifying the heaviness of a beer is not equivalent to specifying its alcoholic strength since the latter depends not only on the content of extract (as measured by gravity) but on the degree of fermentation as well. The following empirical relation holds approximately for gravities in the neighbourhood of 1.045:

$$\text{Alcoholic content} = 0.425 \text{ (degree of fermentation)} \frac{\text{gravity} - 1}{0.004}$$
$$= 106.25 \text{ (degree of fermentation) (gravity)} - 106.25.$$

Since the degree of fermentation is not the same in the various ingredients, there is not a one-to-one correspondence between gravity and alcoholic strength.

[2] *Exercise:* What is the precise interpretation of x_6 and x_7?

of beer is due to this substance. The limits on the content of hop resin can be expressed as the linear restrictions

$$x_{10} + x_{11} = 500 \quad (4\text{g})$$

$$0\,x_1 + 30\,x_2 + 20\,x_3 + 28\,x_4 + 30\,x_5 \quad + x_{11} = 2{,}500 \quad (4\text{h})$$

where x_{10} and x_{11} are slack variables. Of course, the taste of a beer cannot be described exhaustively in terms of content of bitter substance; there are many other taste qualities, but they are difficult to define, let alone to measure objectively[1]. Until further research has been done, the best approach to this problem is to find the solution which minimizes the cost function subject to the restrictions (4a—h) only[2]; if the taste quality of that particular blend is not satisfactory, the "second-best" solution is computed[3], and so forth, until a combination of ingredients is found which is judged satisfactory by experienced tasters. The brewery is not prepared to compromise on the quality of the product; quality is given absolute preference to cost considerations and the problem is to find the least-cost combination of ingredients for a high-quality product.

The coefficients of the mathematical problem are displayed in the following table:

Restric-tions	Variables											Specifications
	x_1	x_2	x_3	x_4	x_5	x_6	x_7	x_8	x_9	x_{10}	x_{11}	
(4 a)	1	1	1	1	1							= 100
(4 b)		2.5	3.7	4.5	5.8							= 310
(4 c)						1	1					= 0.6
(4 d)	1	1.030	1.043	1.050	1.064		1					= 104
(4 e)								1	1			= 200
(4 f)		11	9	8	7				1			= 1,000
(4 g)										1	1	= 500
(4 h)		30	20	28	30						1	= 2,500
		44	50	64	90							$= c = $ min.

Before solving the problem let us examine whether some of the restrictions are redundant. For example, it is conceivable that any non-negative

---

[1] One way of tackling the problem might be to construct a set of grading scales — based upon experiments with a team of beer-tasters — by which the various beer types could be classified with respect to each taste component separately. The gradings would be based on subjective judgment but they would nevertheless be quantitative. Only experiment can show whether it is possible to establish, for each quality component, an order of preference on which most of the tasters agree, and whether a numerical grading scale based upon it will blend linearly.

[2] Only such ingredients are considered as might be expected to give a satisfactory product.

[3] The procedure used for computing second-best solutions is shown in Ch. V, C, below.

solution which satisfies the "strong" restriction (4b) will automatically satisfy the weaker conditions (4c—d). It can easily be proved that this will be the case if

$$\frac{3.1}{1.040} \leq \frac{a_j}{g_j} \leq \frac{3.1}{1.034} \text{ for } j=1, 2, 3, 4, 5 \tag{5}$$

where a_j is the alcohol coefficient and g_j is the gravity coefficient of ingredient no. j (3.1 being the specified alcohol percentage in the mix and 1.034 and 1.040 being the limits on gravity before fermentation)[1]. However, if the actual numerical values of a_j and g_j are inserted, we will find that (5) is not satisfied (as it would have been if the g_j had been nearly proportional to the a_j or the limits had been further apart). In a similar manner it can be shown that inequalities (4e—f) and (4g—h) are not automatically satisfied[2]. Consequently all of the eight restrictions (4a—h) will have to be taken into account and we must look for a solution with (at most) eight positive variables.

An idea which immediately suggests itself is to examine whether the desired product can be made by blending the two cheapest beer types, no. 2 and no. 3[3]. A glance at the coefficients suggests that this may be possible, and the solution for x_2 and x_3 which we can find from (4a—b) actually turns out to satisfy the inequalities (4c—h), the solution being $x_2 = 50$, $x_3 = 50$, $x_6 = 0.25$, $x_7 = 0.35$, $x_8 = 200$, $x_9 = 0$, $x_{10} = 500$, and $x_{11} = 0$[4]. The corresponding value of the cost function is $c = 4,700$ (Danish crowns per 100 hl).

However, it is conceivable that a better solution can be obtained by using water as a third ingredient. Geometrical analysis of the problem in an (x_2, x_3) diagram for $x_4 = x_5 = 0$ will show that the combination is feasible; indeed there will be more than one such solution since, for $x_4 = x_5 = 0$, we have eight equations in nine variables[5]. One such combina-

---

1 *Exercise:* Prove this proposition.

2 *Exercise:* Carry out the calculations.

3 They are not only cheaper *per hl* than nos. 4 and 5. Dividing the coefficients in each column by the corresponding price we will find that ingredient no. 3 is the cheapest source of alcohol while no. 2 gives the largest contribution to gravity, colour, and hop resin content per crown's worth. (Cf. the ice cream problem of Ch. III.)

4 This is a "degenerate" solution with fewer than eight positive variables.

5 Although there are three structural variables, geometrical illustration in two dimensions is possible if x_1 is treated as a slack variable in (4a) and replaced by $(100 - x_2 - x_3)$ in (4d). For $x_4 = x_5 = 0$, (4a) represents a half plane in an (x_2, x_3) diagram, whereas (4b) corresponds to a straight line. The three pairs of upper and lower limits, cf. (4c–h), are represented by three belts in the (x_2, x_3) plane. These restrictions together define a set of feasible solutions.

Exercise: Draw the diagram on graph paper and show the positions of the basic solutions that exist for x_2 and $x_3 > 0$ and $x_4 = x_5 = 0$. How many are there and how many of them are feasible?

tion (basic feasible solution) is of course the solution above where $x_1=0$, another is that in which $x_{10}=0$. The latter solution is $x_1=9.84$, $x_2=19.67$. $x_3=70.49$ and total cost of ingredients will be $c=4{,}389.98$, which is a cheaper solution than the one we found above. To examine whether it is an optimal solution, let us solve for the eight basic variables in terms of the non-basic variables x_4, x_5, and x_{10}. This can be done very easily because of the many zeroes among the coefficients of the equation system[1]. The solution is

$$\begin{aligned} x_1 &= 9.84 + 0.29\,x_4 + 0.54\,x_5 - 0.02\,x_{10} \\ x_2 &= 19.67 - 0.22\,x_4 + 0.08\,x_5 + 0.06\,x_{10} \\ x_3 &= 70.49 - 1.07\,x_4 - 1.62\,x_5 - 0.04\,x_{10} \\ x_6 &= 0.22 - 0.01\,x_4 - 0.01\,x_5 + 0.00\,x_{10} \\ x_7 &= 0.38 + 0.01\,x_4 + 0.01\,x_5 - 0.00\,x_{10} \\ x_8 &= 50.78 - 4.05\,x_4 - 6.70\,x_5 + 0.30\,x_{10} \\ x_9 &= 149.22 + 4.05\,x_4 + 6.70\,x_5 - 0.30\,x_{10} \\ x_{11} &= 500.00 \qquad\qquad\qquad\qquad - 1.00\,x_{10} \end{aligned} \qquad (6)$$

and substituting in the preference function we get

$$c = 4{,}389.98 + 0.82\,x_4 + 12.52\,x_5 + 0.64\,x_{10}.$$

The simplex criterion is satisfied, so the corresponding basic feasible solution — as represented by the constant terms — is optimal. Thus we have solved the problem in one step and the least-cost combination is to mix 9.84 hl of water with 19.67 and 70.49 hl of the first two types of beer. The solutions for the slack variables indicate that the average gravity before fermentation is 1.036 whereas the colour index of the blend will be 8.51 and the content of hop resin will be 20 mg per litre. Provided that this blend turns out to have a satisfactory taste, the solution represents the best course open to the brewery. If, however, this particular blend is rejected by the brewery's tasters, other basic feasible solutions will have to be computed and examined for taste quality (viz., the solutions found by introducing x_4, x_4, or x_{10} into the basis).

B. Optimal Utilization of Machine Capacities

1. In the mechanical industry, the following type of optimization problem is frequently met with: A company manufactures a number of products which have to be processed on the same machines whose capacities set limits to production per period. The problem is to determine the

---

[1] For example, (4b), (4g), and (4h) can be solved for x_2 and x_3; (4a) then gives x_1, and so forth.

quantity to be produced of each product[1] when total profit is to be maximized subject to these capacity limitations, i. e., to find a feasible production programme which represents *optimal utilization of the company's fixed equipment*. This is evidently a linear programming problem when the usual conditions of linearity are satisfied.

The following hypothetical example[2] will illustrate the problem. An automobile manufacturing plant makes two distinct products, trucks and automobiles. The plant is organized in four departments each of which has a limited capacity. The first is the sheet metal stamping department, whose capacity can be expressed as either 35,000 trucks of 25,000 automobiles per month. Thus one truck occupies 1/35,000 of the monthly capacity whereas one automobile requires 1/25,000, and the capacity limitation can be written as the linear inequality

$$\frac{1}{35,000} x_1 + \frac{1}{25,000} x_2 \leq 1 \qquad (7)$$

where x_1 and x_2 are the numbers of trucks and automobiles produced per month[3]. The second department is the engine assembly department, whose capacity is 16,667 trucks or 33,333 automobiles per month. Third, the capacity of the automobile final assembly department is 22,500 automobiles per month, whereas trucks are assembled in a separate fourth department with a monthly capacity of 15,000. Thus we have the linear programming problem[4]:

		Product no.		
		1	2	
"Machine"	1	1/35,000	1/25,000	≤ 1
(Department)	2	1/16,667	1/33,333	≤ 1
no.	3		1/22,500	≤ 1
	4	1/15,000		≤ 1
		250	300	$= f =$ maximum

where the coefficients of the profit function, $ 250 and 300, are selling price minus variable cost per unit of each product; the costs associated

---

1 This includes the decision as to which products are to be made at all.

2 Borrowed from R. Dorfman (1953).

3 Alternatively, the capacity restriction may be written with machine hours per truck and per automobile as coefficients on the left-hand side and machine hours available per month on the right-hand side. This form of the restriction is obtained from (7) by multiplying by the number of machine hours available per month.

4 *Exercise*: Solve the problem geometrically to find the optimal basis, and compute the exact numerical solution.

with the use of the fixed equipment can be disregarded since they are fixed charges which cannot affect the position of the optimal solution. The problem is evidently quite analogous with the second type of blending problem above (the allocation of scarce raw materials between several products)[1]. Each product defines an activity and there is a restriction for each scarce or fixed factor of production[2].

2. Some problems of optimal capacity utilization have additional restrictions in the form of *upper or lower limits on the x's*. Sales potentials may be limited, or orders received may set lower limits to the production programme. A further generalization of the model is the case of *more than one activity (method of production) for each output*.

For example, multiple shift operation can be handled within the framework of a linear model by introducing separate activities for the outputs of the two or three shifts; thus the variables of the problem will be x_{ij} where the subscripts denote shift number and product number respectively. The technical coefficients of the several activities associated with the same product may be identical, but the coefficients in the linear profit function will be different because labour cost per unit of product is higher in the second and third shifts. The case of overtime work can be treated similarly, normal and overtime output being produced in technologically identical but economically distinct processes.

Another example is the case where the machines are, to a certain extent, capable of doing each other's jobs. For instance, certain operations may be performed either by a lathe or by a milling machine; or the amount of work to be done by a grinder can be reduced by closer machining[3], i.e., lathe time is substituted for grinder time. These cases are genuine examples of input or factor substitution, the existence of several processes (sets of technical coefficients) for each product implying that the same output can be produced by several different combinations of the services of the fixed factors of production.

The following concrete example[4] is characterized by this type of factor substitution. A company produces six products using eight different types of screw machines and three types of grinders. The capacities of these machine groups give rise to eleven restrictions, in addition to which there are six (minimal) production requirements. The number of structural activities is 34; each product can be made in several alternative processes,

---
1 Cf. pp. 30—37 above.
2 The only difference is that, while it may sometimes be possible to sell unused amounts of raw materials at a positive price, the slack variables in a machine capacity problem will almost always have zero coefficients in the profit function since idle machine hours can rarely be "sold" (in the form of letting out the machines).
3 Cf. A. Henderson and R. Schlaifer, op. cit., p. 87.
4 Borrowed from A. Charnes, W. W. Cooper, D. Farr, and Staff (1953).

i.e., by several different combinations of screw machine and grinder time[1]. A characteristic section of the complete table of coefficients is given in the table below, where the product quantities — i.e. the structural variables — are measured in thousands of units per period, the technical input coefficients in machine hours per 1,000 units of product, and the profit coefficients in $ per 1,000 units. $P_1, P_2, \ldots$ denote the activities (processes).

		Product no. 1				... 5				...	Specifications
		P_1	P_2	P_3	P_4	...	P_{16}	P_{17}	P_{18}	P_{19}	...	
Screw machine type no.	1	2.23					171.27	171.27				$\leq$ 942
	2		2.49									$\leq$ 1,050
	3			5.25								$\leq$ 2,280
	4				5.04							$\leq$ 2,127
	5											$\leq$ 68
	6											$\leq$ 978
	7											$\leq$ 553
	8								196.80	196.80		$\leq$ 1,115
Grinder type no.	1						18.58		18.58			$\leq$ 105
	2											$\leq$ 12
	3							123.98		123.98		$\leq$ 702
Product no.	1	1	1	1	1							$\geq$ 422.0
	2											$\geq$ 117.0
	3											$\geq$ 200.0
	4											$\geq$ 3.5
	5						1	1	1	1		$\geq$ 5.5
	6											$\geq$ 9.0
Profit, $ per 1,000 units		212.5					21,142.9					$=f=$ max.

A large company in the ball bearing industry (SKF Industries, Inc.) is reported to have saved about $ 100,000 per year by applying linear programming techniques to a production planning problem of a similar type[2].

---

1 The actual number of products was far greater than six, but they were arranged in six major groups in order to simplify the problem. Similarly, each of the eleven machine groups of the problem represents a number of different machine types which have been reduced to one type by applying suitable correction factors (efficiency factors).

This aggregation procedure has since been abandoned as it proved possible to solve the complete problem numerically without first having to simplify it in this fashion. Cf. op. cit., p. 117.

2 Cf. A. Henderson and R. Schlaifer, op. cit., p. 86.

When the production requirements in problems of this type are specified on an exact basis, i.e., when they are equations rather than inequalities, total gross revenue from sales becomes a constant which can be disregarded so that we have a problem of allocating the machines such as to minimize total variable cost[1].

3. Problems of optimal utilization of fixed equipment will often be complicated by the fact that the machine capacities — machine hours available per period — have to be adjusted for *setup times*. This correction cannot be made beforehand, because total setup time for each machine will depend on the choice of products to be made and the choice of processes to be applied, i.e., on the production programme which has yet to be determined. On the other hand it is not possible to build the correction into the linear model, as setup time for a particular operation[2] is not proportional to the level of output; it will be a constant, independent of the activity level, if the process is used at all, and zero if it is not applied (i.e., if the corresponding variable turns out to be zero in the optimal programme).

In many cases this complication can, of course, be disregarded because setup times are insignificant in relation to available machine hours, as they will be when the planning horizon is long enough for the products to be made in long runs. If that is not the case, however, the problem raised by setup times can be handled by an iterative procedure. First, the setup times are estimated and the capacities adjusted accordingly. The second step is to compute a solution which will be optimal provided that setup times were estimated correctly. Third, the setup times that this programme would actually involve are compared with the original estimates; if the discrepancy is considered significant, the capacities are adjusted in accordance with the programme and a new solution which is optimal with respect to the readjusted capacity limitations is computed — and so forth, until a programme has been reached where the "actual" setup times are in agreement with those on which the calculations are based[3]. Evidently this procedure will be considerably shortened if the first estimates of setup times are substantially correct.

4. Another complication is the *scheduling* problem. The solution to a programming problem concerned with optimal utilization of given capaci-

---

1 An example of this "machine assignment problem" is found in A. Henderson and and R. Schlaifer, op. cit., pp. 89f. It differs from the example above in that any of the machines can produce any of the products, i.e., there is an activity for each combination of a machine and a product and each activity has only one input coefficient in it.

2 I.e., setup time on a particular machine for a particular activity.

3 See A. Henderson and R. Schlaifer, op. cit., pp. 86–89, where the procedure is illustrated by a numerical example.

ties tells us what operations are to be performed and on which machines, but it does not tell us anything about the timing of the operations, i.e., the order in which the jobs are to be processed on each machine. Hence the calculation of the optimal programme will have to be followed by the determination of a sequence of operations on each machine which will schedule the operations through the plant at minimum idle time, or whatever criterion is selected for schedule evaluation. However, when each product has to be processed on the various machines in a technologically given sequence, it may not be possible at all to schedule the otherwise optimal production programme through the plant without violating the capacity limitations; idle time between operations cannot be avoided because certain operations must wait till others have been finished, so the programme may require more machine hours — working or idle — than there are available within the period for which production is planned. In such cases the optimal programme will have to be modified.

Consider, for example, the problem

$$1.5\,x_1 + 5\,x_2 \leq 40 \quad \text{(machine no. 1)}$$
$$2\,x_1 + 4\,x_2 \leq 40 \quad \text{(machine no. 2)}$$
$$2\,x_1 + 5\,x_2 = f = \text{maximum}$$

where x_1 and x_2 are quantities produced of two commodities on two machines; the planning period is a week (40 hours). The optimal solution is

$$x_1 = 10, \quad x_2 = 5 \quad \text{(units/week)},$$

which is seen to require exactly 40 machine hours per week on each machine.

The next step, then, is to set up a *schedule* of operations, an operation being defined as the processing of product no. j on machine no. i. The *processing times* of the operations (a_j and b_j respectively on the two machines) are shown in the following table, where each product is considered as a *job* requiring time on both machines.

Processing times (hours)		Machine no. $i=$	
		1	2
Job (product)	1	$a_1 = 1.5\,x_1 = 15$	$b_1 = 2\,x_1 = 20$
no. $j=$	2	$a_2 = 5\,x_2 \;\;= 25$	$b_2 = 4\,x_2 = 20$
Total		$A = a_1 + a_2 = 40$	$B = b_1 + b_2 = 40$

Now if there are no restrictions on the order of operations — one machine does not have to wait till the other has completed its part of the job — the scheduling problem is simple enough. The operations on each machine can be performed in arbitrary order without any intermediate waiting time, and the *total flow time* from the start of the first operation to completion of the last will be 40 hours so that the programme can be carried out within the planning period.

In practice, however, there may be technological restrictions on the order in which each job (product) is to be processed on the machines, e.g., first on a lathe, second on a grinder. This means that the second operation of a job cannot start until the first operation has been completed, and the resulting waiting time will lengthen the total flow time. The planner is then confronted with the problem of determining a schedule which *minimizes total flow time*[1].

Let us consider the example above, specifying the assumptions as follows:

(i) Each machine can work on only one job at a time; i.e., a_1 cannot overlap a_2 in time, and b_1 cannot overlap b_2;

(ii) each job can be processed on only one machine at a time, that is, a_j and b_j cannot overlap $(j=1, 2)$;

(iii) each job is to be processed first on machine no. 1, second on machine no. 2, so that a_j is to precede b_j $(j=1, 2)$.

However, we are still free to select *the order in which the two jobs are processed on each machine*. There are two possible sequences on the first machine: *1,2* (job no. 1 first, job no. 2 second, i.e., a_1 precedes a_2) and *2,1*, and similarly on the second machine so that we have the following four combinations:

Sequences		1°	2°	3°	4°
Machine no.	1	1,2	2,1	1,2	2,1
	2	1,2	2,1	2,1	1,2

For each combination we can determine a schedule which meets the requirements (i)—(iii), using a *Gantt chart* where the processing times are arranged on parallel time axes, one for each machine.

The Gantt chart of *combination 1°* is shown in Fig. 4[2]:

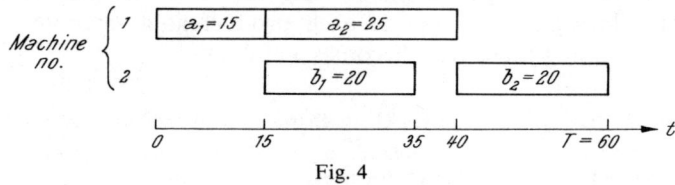

Fig. 4

---

[1] Minimum flow time is only one of many possible criteria of optimality in scheduling problems. See, e.g., R. W. Conway, W. L. Maxwell, and L. W. Miller (1967).

[2] Of course, b_1 might as well start five hours later, thus doing away with the idle time interval between b_1 and b_2.

Total flow time is seen to be $T=60$ hours. *Combination* $2°$ leads to the schedule shown in Fig. 5, with flow time $T=65$ hours:

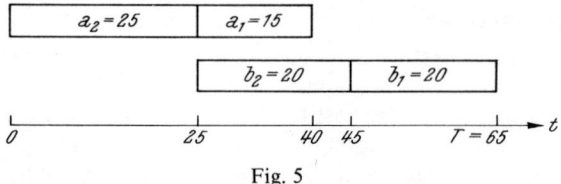

Fig. 5

Combinations $3°$ *and* $4°$ are both seen to imply that both operations on one machine will have to be completed before the other machine can start, so that total flow time will be $T=A+B=80$ hours. Obviously the situation can be improved by reversing the order of operations on one of the machines, which leads to either $1°$ or $2°$ [1]. Consequently we need only consider cases where the jobs are performed in the same order on the two machines[2], that is, the possible *sequences of job numbers* (*1,2* and *2,1*), of which the sequence *1,2* (combination $1°$, Fig. 4) is seen to represent minimum flow time in the example, $T=60$[3].

This schedule, however, cannot be implemented within a calendar week of 40 hours, although it is based on a feasible solution to the linear programming problem. The reason is, of course, that it was not specified in the programming problem that the two capacity restrictions (40 hours available on each machine) should refer to the *same* calendar period (the "same" 40 hours).

Awkward as the situation seems, there are two possible ways of evading the dilemma.

In the first place, we have tacitly assumed that the second machine will have to wait for *the whole lot* of each product to be completed on the first machine. This may not always be so. If it is technologically possible to transfer *the individual unit* of product to the second machine as soon as it is finished on the first — which means that each unit of product can be considered as a separate job — the problem becomes much less important. Treating each unit of each product as a separate job, the table of processing times now becomes as follows:

---

1 *Exercise:* Show this by drawing Gantt charts of combinations $3°$ and $4°$.

2 This result can be shown to hold for any number of jobs to be processed on two machines (but not for any number of machines).

3 For larger numbers of jobs, graphical solution of the scheduling problem by Gantt charts becomes impracticable. For $n=6$, the number of possible sequences (permutations of job numbers), and therefore the number of charts to be drawn, will be $6!=720$. In such cases numerical methods must be resorted to, for example, *Johnson's algorithm* (cf. R. W. Conway *et al.* (1967), Ch. 5).

Processing times (hours)		Machine no. $i=$	
Product no.	Job no. $j=$	1	2
1	1—10	$a_j=1.5$	$b_j=2$
2	11—15	$a_j=5$	$b_j=4$
Total		40	40

Processing these fifteen jobs in the order of job numbers gives a schedule with a total flow time of $T=44$ hours[1] as against $T=60$ above; thus the excess time requirement has been reduced from 50 to 10 per cent. Clearly, the larger the production programme is in proportion to the machine time coefficients, the closer we get to the ideal situation where the programme can be carried out within the 40-hour period.

Second, even if the lots to be produced per week are indivisible jobs, the fact that production plans are often made for a number of successive identical periods will reduce the excess problem. Suppose that production is planned on a weekly basis for four weeks; the programme for each week is assumed to be as above and a week's production constitutes an indivisible job. We can then draw a Gantt chart for a four-week horizon, repeating the optimal weekly schedule 1,2 as shown in Fig. 6:

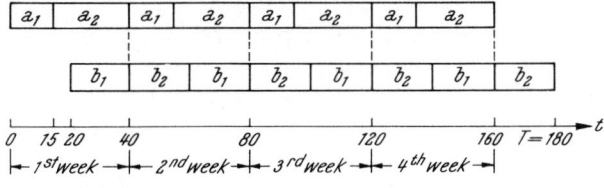

Fig. 6

The absolute excess time requirement is still 20 hours (as in Fig. 4), but it has become much less important in proportion to the total calendar time available $(20/160=12.5$ per cent as against 50 per cent in the weekly scheduling problem). The longer the horizon is in proportion to the planning period within which the production lot is indivisible, the more we approach the ideal situation where the trouble caused by the scheduling problem is negligible and the optimal solution to the linear programming problem can be carried out without having to be revised.

C. Inventory Problems

In the previous examples we have disregarded inventories, tacitly assuming that what is produced is also sold (and that raw materials are purchased for immediate use). Clearly this is not always a realistic as-

---
[1] *Exercise:* Demonstrate this by sketching the Gantt chart.

sumption; in many cases — particularly when the demand for the company's products is subject to fluctuations — the time rates of production and sales may be different because the rate of production can be smoothed by allowing inventories to fluctuate. Thus, when the time pattern of expected sales is given, *the problem of optimal production planning implicity becomes a problem of planning inventories*; the difference between the time rates of production and sales represents the net increase in inventory per unit of time, and the size of the inventory at any time will be equal to initial inventory (at the beginning of the planning horizon) plus cumulated production minus cumulated sales.

Inventory variations will of course be of particular importance in the case of seasonal or other marked fluctuations in expected sales during the planning period (e.g., during the year). When the cost of producing a unit of output increases with the quantity produced, for example, because of the higher labour cost associated with overtime or night-shift production, this fact in itself calls for a *uniform rate of production* during the period since this will lead to the lowest possible cost of producing the total programme. Against this we have to consider the cost of carrying the resulting inventories; from the point of view of minimizing storage costs the best solution is to *synchronize production with sales* as far as possible. The optimal programme — the production schedule that meets the sales requirements at minimum total cost of production and storage—will clearly have to be some sort of a compromise between these two extremes. In spite of the dynamic element which is brought into the picture by inventory fluctuations, problems of this nature can be solved by linear programming methods, as the following concrete example will show[1].

After analyzing the market conditions, a tobacco factory has drawn up a sales program for a particular cigar brand, covering a total period of one year. According to the program, estimated sales (in thousands of cigars) in each of the four quarters are as follows:

Quarter no.	Expected sales, thousands	Cumulated sales, thousands
1	712	712
2	578	1,290
3	547	1,837
4	1,216	3,053

For the year as a whole the company is planning a total production equal to total expected sales, initial and terminal inventories being assumed to

---

[1] The data were kindly provided by the tobacco factory of the Danish Cooperative Union. The model was developed in collaboration with Professor E. Lykke Jensen, Copenhagen.

be zero[1], and the problem is to determine production in each of the four quarters in such a way as to minimize total cost of production and storage. The factory operates in two shifts with the capacity limits 550.5 and 453.0 respectively[2] (in thousands of cigars per quarter). The cost of producing 1,000 cigars is 60.62 Danish crowns in shift 1 and 73.66 D. cr. in the second shift; only labour costs are considered, all other cost items being fixed in relation to the present problem. The cost of storing 1,000 cigars for a quarter, exclusive of fixed charges, is 1.01 D. cr. Let x_{it} denote production in shift no. i in the tth quarter ($i=1, 2$; $t=1, 2, 3, 4$). Then the problem is to minimize total variable cost subject to the restrictions

$$\begin{aligned}
x_{11} & & +x_{21} & & &\geq 712 \\
x_{11}+x_{12} & & +x_{21}+x_{22} & & &\geq 1{,}290 \\
x_{11}+x_{12}+x_{13} & & +x_{21}+x_{22}+x_{23} & & &\geq 1{,}837 \\
x_{11}+x_{12}+x_{13}+x_{14} & +x_{21}+x_{22}+x_{23}+x_{24} & & & &= 3{,}053 \\
x_{11} & & & & &\leq 550.5 \\
x_{12} & & & & &\leq 550.5 \\
x_{13} & & & & &\leq 550.5 \\
x_{14} & & & & &\leq 550.5 \\
x_{21} & & & & &\leq 453 \\
x_{22} & & & & &\leq 453 \\
x_{23} & & & & &\leq 453 \\
x_{24} & & & & &\leq 453,
\end{aligned}$$

of which the first four express the requirement that inventory must be non-negative at the end of each quarter (i.e., that delays in meeting demand are not admissible) and zero at the end of the year, whereas the last eight inequalities are the capacity restrictions[3]. Total cost of production for the year will be a linear function of the same variables,

$$c_1 = 60.62\,(x_{11}+x_{12}+x_{13}+x_{14}) + 73.66\,(x_{21}+x_{22}+x_{23}+x_{24})\ (\text{D.cr.}).$$

Thus the apparent nonlinearity of the cost function has been overcome by the simple device of introducing separate variables and separate capacity limitations for the two shifts, i.e., for the two linear segments of which the cost function for each quarter is omposed. Storage cost in each quarter is 1.01 times the average number of units (thousands of cigars) held in inventory during the quarter, as approximated by the

---

1 In actual practice, a positive inventory will usually be left over from the previous period, even if it was intended to close that period with zero stock. However, the general nature of the model will not be changed if a positive inventory is entered at the beginning and at the end of the planning period.

2 These are the capacities of the roll-making machines which are the bottleneck factors of the planning problem.

3 *Exercise:* Interpret the slack variables in all of the 11 inequalities.

mean of initial and terminal inventories; thus for the first quarter we get

$$1.01 \frac{0+(x_{11}+x_{21}-712)}{2}$$

and for the entire year

$$c_2 = 1.01\,(3\,x_{11} + 2\,x_{12} + x_{13} + 3\,x_{21} + 2\,x_{22} + x_{23} - 3{,}839)\ (\text{D. cr.}).$$

The complete cost expression then becomes the linear function

$$c = c_1 + c_2 = 63.65\,x_{11} + 62.64\,x_{12} + 61.63\,x_{13} + 60.62\,x_{14}$$
$$+ 76.69\,x_{21} + 75.68\,x_{22} + 74.67\,x_{23} + 73.66\,x_{24} - 3{,}877.$$

The optimal solution can be computed by the simplex method (neglecting, of course, the constant term in the cost function)[1]. However, the structure of the present type of model is simple enough to permit solution by inspection, observing the following two self-evident rules of thumb: (i) in each quarter, nothing should be produced in the second shift as long as there is idle capacity in the first shift in the same quarter, nor (ii) as long as there is idle capacity in the first shift in any previous quarter *and* the cost of storing a unit is less than the difference in cost of production per unit between the two shifts (i.e., 1.01 (storing time) < 73.66 − 60.62, an inequality which in this case will always be satisfied since storing time is less than 4 quarters)[2].

The following table, where s_t denotes expected sales and x_t is total planned production ($= x_{1t} + x_{2t}$) in the tth quarter, shows how the problem is solved by this method.

t	s_t	$x_t\,(\leq 1{,}003.5)$	of which	
			$x_{1t}\,(\leq 550.5)$	$x_{2t}\,(\leq 453.0)$
1	712	712.0	550.5	161.5
2	578	578.0	550.5	27.5
3	547	547.0 + 3.5 + 209.0	547.0 + 3.5	0.0 + 209.0
4	1,216	1,003.5	550.5	453.0
Total	3,053	2,840.5 + 3.5 + 209.0	2,198.5 + 3.5	642.0 + 209.0

The two first quarters do not present any problems. In the third quarter, however, the sales require less than full utilization of the capacity of shift 1, while the reverse is the case with the fourth quarter. Now, by rule (ii), we should first use up the idle capacity in the third quarter

---

[1] The solution is: $x_{11} = x_{12} = x_{13} = x_{14} = 550.5$, $x_{21} = 161.5$, $x_{22} = 27.5$, $x_{23} = 209.0$, $x_{24} = 453.0$, and $c = 196{,}385$.

[2] Storing here implies that unutilized capacity in the first shift is transferred to following quarters.

(3.5 units in shift 1) before operating shift 2 in the fourth quarter, producing 453.0 units. We are still 209.0 units short, which have to be produced in shift 2 in the third quarter. In this way we get $x_3 = 547.0 + 3.5 + 209.0 = 759.5$ and $x_4 = 1,003.5$.

This model is, of course, only a rough sketch of the actual problem; seasonal fluctuations cannot be handled adequately by dividing the year into four sub-periods only. We will get a much more detailed picture if the planning period is considered as composed of twelve months, in accordance with company practice[1].

In many practical cases of this nature further restrictions will have to be taken into consideration. Storing capacity may be limited; or it may be prescribed that the product should not be stored for more than a limited period of time, as will be the case with perishable commodities. These additional side conditions can be expressed as linear inequalities in the variables of the model, so we still have a linear programming problem[2]. Another generalization of the model is the introduction of joint products (for example, several cigar brands made on the same roll-making machines) which are interdependent because they have to share the limited capacities. In this case, too, the model remains linear and the problem can be solved by means of a slightly modified version of the above inspection method[3].

A somewhat related type of problem is the following "warehousing problem"[4]: A company produces (or purchases) and sells a product whose price is subject to know seasonal fluctuations. Inventory (warehousing) capacity is limited. Initial inventory being given, the problem is to determine the optimal time patterns of purchases, inventories, and sales

---

1 *Exercise:* Solve the problem in this form (and with three shifts instead of two), using the inspection method described above. The data are as follows:

$t =$	1	2	3	4	5	6	7	8	9	10	11	12	Total
$s_t =$	209	164	339	203	183	192	198	181	168	596	518	102	3,053

The monthly capacities for the three shifts are 183.5, 151.0, and 122.3 respectively. Cost of production per unit in the third shift is 90.93 (D. cr. per 1,000 cigars) and storage cost is 0.34 D. cr. per unit per month.

2 In a graph showing cumulated production (according to the solution) and cumulated sales as functions of t, storing time and inventory are represented respectively by the horizontal and vertical distances between the curves at time t, and we can then examine whether the solution respects the additional inequalities. If this is not the case, the inspection methods breaks down and the production program will have to be recomputed by the simplex method, taking explicit account of *all* restrictions on the variables.

Exercise: Indicate the general form of the additional constraints.

3 For an example, see S. Danø and E. L. Jensen (1958).
4 Cf. A. Charnes and W. W. Cooper (1955).

over the year (i.e., for each month or quarter). This problem can be solved by linear programming, and it is readily extended to cover the case of several products and several warehouses.

D. Transportation Problems

1. The next type of problem which we shall demonstrate is the so-called *transportation problem*.

When a company's activities involve freight shipments between *geographically separated departments* — for example, between plants and warehouses located in different parts of the country — it becomes an important problem to schedule the shipments in such a way as to minimize total cost of transportation. This problem can in many cases be formulated and solved as a linear programming problem.

Suppose that a company produces the same commodity in two factories, one in the East and one on the West Coast, and distributes the product from three warehouses scattered across the country. Total production per week, 600 tons, is to be distributed among the warehouses, the two factories producing 200 and 400 tons and the warehouses requiring 300, 100, and 200 tons per week respectively, and the problem is *to determine an optimal pattern of shipments*, i.e., to determine the amounts to be shipped from each plant to each warehouse per week when total cost of transportation is to be as low as possible.

The unknowns of the problems are shown in the following table, where x_{ij} denotes the amount (in tons per week) supplied from factory no. i to warehouse no. j.

x_{ij}, tons per week		Warehouse no.			Total
		1	2	3	(production)
Factory no.	1	x_{11}	x_{12}	x_{13}	200
	2	x_{21}	x_{22}	x_{23}	400
Total (requirement)		300	100	200	600

There are $2 \times 3 = 6$ such unknowns, each corresponding to a possible route of transportation, as shown in Fig. 7.

Each variable has a cost coefficient c_{ij} associated with it, denoting the cost of shipping one ton from factory no. i to warehouse no. j. These freight rates are given in the following table.

c_{ij}, $ per ton		Warehouse no.		
		1	2	3
Factory no.	1	3	2	1
	2	2	4	7

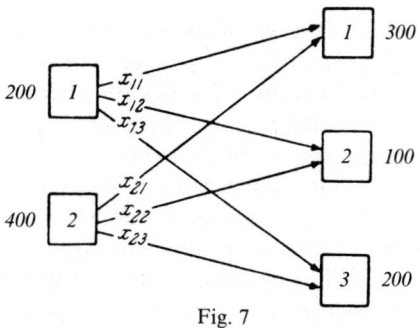

Fig. 7

The problem is to minimize total cost of transportation,

$$c = \sum_i \sum_j c_{ij} x_{ij},$$

subject to the conditions that total shipments from the two factories shall be 200 and 400 and that total deliveries to the three warehouses shall be 300, 100, and 200 tons per week respectively (i.e., that the row sums and the column sums in the first table shall be as indicated in the margin). Since c is a linear function of the x_{ij} and the side conditions are mere summations of these variables, this is a linear programming problem[1]:

$$\begin{aligned}
x_{11} + x_{12} + x_{13} &= 200 \\
x_{21} + x_{22} + x_{23} &= 400 \\
x_{11} \qquad\qquad + x_{21} &= 300 \\
x_{12} \qquad\qquad + x_{22} &= 100 \\
x_{13} \qquad\qquad + x_{23} &= 200
\end{aligned}$$

$$3x_{11} + 2x_{12} + 1x_{13} + 2x_{21} + 4x_{22} + 7x_{23} = c = \text{minimum},$$

indeed, a particularly simple one since the coefficients of the x_{ij} are either 0 or 1, arranged in a regular pattern[2].

A glance at the example shows that each of the five equations can be derived from the four others, so that there are only four independent restrictions; this follows from the fact that we get the same results by adding the first two equations (the "factory equations") as we get if we add the three "warehouse equations", namely, that the sum of all of the variables is equal to 600, the total quantity to be shipped. The same result holds, of course, for any problem of this type; in the general case of m origins and n destinations, one of the $m+n$ equations can — and should — be left out, no matter which one, and it takes only $m+n-1$ variables to form a basis. It follows that the optimal transportation

---

[1] *Exercise:* Why are the variables required to be non-negative?

[2] This fact makes it possible to solve such problems by a considerably simplified variant of the simplex method, see Ch. V, F.

pattern will make use of at most $m+n-1$ of the $m \cdot n$ possible routes[1]. It is difficult to see how this result could have been arrived at without formulating the problem in terms of a linear programming model and using the fundamental theorem[2].

Transportation planning is one of the fields where linear programming methods have been applied most successfully. It is reported[3], for example, that the H. J. Heinz Co., which manufactures tomato ketchup in half a dozen factories and distributes the product to about 70 warehouses, has saved several thousand dollars semiannually in freight costs by applying linear programming to determine the optimal pattern of shipments. Furthermore, the great simplicity of the method has made it possible for the company to revise the optimal transportation programme at shorter intervals and thus to obtain a better adjustment to changes in the data of the problem.

The transportation model covers a wide variety of practical planning problems[4]. During the Second World War, the Allies were confronted with the problem of scheduling transatlantic shipments in such a way as to minimize total ballast time (although the linear programming methods for solving such problems were not developed until after the war), and more or less similar problems have to be solved by large companies in the transportation industries, e.g., shipping companies, railways, etc.

2. The numerical example above is a *balanced* transportation problem: the restrictions have the form of *equations*, and total shipments from the two factories (200+400) are equal to the total amount delivered to the three warehouses (300+100+200); were this not the case the equations would be inconsistent.

However, in many practical applications the amounts to be shipped from each factory are not specified on an exact basis but are limited by plant capacities so that the restrictions associated with the factories become *inequalities*. Similarly, the amounts to be delived to each warehouse may be specified by upper limits only. In such cases the right-hand sides of the two groups of restrictions cannot be expected to add up to the same total and we have an *unbalanced transportation problem*.

A slightly generalized transportation model has been devised to cover

---

[1] Solutions of transportation problems will often turn out to be degenerate, ie., fewer than $m+n-1$ variables are positive in the optimal combination of routes.

[2] Quite apart from the question of number of routes, it is difficult to solve transportation problems by common-sense methods. It is no use assuming that each warehouse shall be supplied from the nearest factory, for this assumption will only in exceptional cases be compatible with the restrictions.

[3] Cf. A. Henderson and R. Schlaifer, op. cit., pp. 77f.

[4] See, for example, A. Henderson and R. Schlaifer, op. cit., pp. 79ff., or R. L. Ackoff (1961), pp. 139–150.

problems of this type. It is based on the simple trick of introducing fictitious shipments (with zero cost coefficients) such as to turn the problem into a balanced one.

In the example above, let the capacities of the two factories be 300 and 500 tons per week, and let us assume that production in each factory is no longer given but will be limited by the capacity. Total capacity (800) now exceeds total stipulated deliveries to the warehouses (600) so that there will be unutilized capacities, corresponding to slack variables in the factory restrictions. We can now interpret the slack variables — to be denoted by x_{14} and x_{24} — as fictitious deliveries to a fourth, fictitious warehouse ("dummy destination"). By definition they add up to $800-600=200$ so that we now have a balanced transportation table:

x_{ij}, tons per week	Warehouse no.				Total (capacity)
	1	2	3	4	
Factory no. 1	x_{11}	x_{12}	x_{13}	x_{14}	300
2	x_{21}	x_{22}	x_{23}	x_{24}	500
Total (requirement)	300	100	200	200	800

In addition to the costs of transportation, the cost function which is to be minimized will now have to include the cost of production. Total production is still given, but the cost of producing the 600 tons per week will depend on how much of this quantity is allocated to each factory. Let variable unit costs of production be $ 5 and $ 4 per ton respectively. We must then add 5 to the coefficients in the first row and 4 to those in the second row of the cost table above, while the cost coefficients in the new column are set $=0$ because the cost of producing and shipping nothing is zero:

c_{ij}, $ per ton	Warehouse no.			
	1	2	3	4
Factory no. 1	5+3	5+2	5+1	0
2	4+2	4+4	4+7	0

The problem can now be formally handled as a balanced transportation problem (with two factory equations and four warehouse equations) where the *total cost of production and transportation is to be minimized*[1,2].

[1] In contrast to the cost of production, the total sales revenue can be taken as independent of the allocation of production to factories because the two factories produce the same commodity; consequently sales revenue can be neglected. The solution which minimizes total cost will also maximize total profit.

[2] A similar model applies to cases where the problem is to minimize the total cost of purchasing a raw material (or a finished product) from a number of geographically scattered suppliers and having it shipped to a number of differently located plants (or warehouses) which the company owns.

As another example, suppose that three production plants belonging to the same company and producing the same commodity are to serve five distribution centres. The plants are operated at full capacity, producing 100, 70, and 40 tons of the commodity per week. The distribution centres can receive (unload and sell to consumers) up to 50, 45, 40, 75, and 35 tons per week respectively. The shortest road distances c_{ij} (in miles) from plant no. i to centre no. j being given, the problem is to determine the shipments x_{ij} which *minimize the total volume of transportation* as measured in ton-miles per week.

The total production, which is fixed at 210 tons per week, is seen to be less than the total amount the distribution centres can receive (245 tons). To balance the problem we introduce a fourth, fictitious plant (a "dummy origin") with a fictitious production of $245 - 210 = 35$ tons per week, of which x_{4j} tons are "shipped" to distribution centre no. j ($j = 1, 2, 3, 4, 5$); since no real transportation is involved, we can set the distances c_{4j} equal to 0. The x_{4j}, which add up to 35, can be identified with the slack variables of the restrictions corresponding to the distribution centres. All the relevant information is contained in the following table.

c_{ij} (miles)	Distribution centre no.					Total (production, tons per week)
	1	2	3	4	5	
Plant no. 1	14	48	12	35	40	100
2	16	25	28	30	57	70
3	25	32	53	14	23	40
4	0	0	0	0	0	35
Total (requirements, tons per week)	50	45	40	75	35	245

The problem is now ready to be solved as a balanced transportation problem with $4 + 5 - 1 = 8$ independent restrictions in equality form.

Still more general cases exist where neither deliveries to the warehouses (i.e., the geographical distribution of sales) nor shipments from factories (the allocation of production to plants) are specified exactly. This means that slack variables have to be introduced into all of the restrictions in order to turn the problem into a balanced one, and that the optimal solution is found by *maximizing sales revenue minus variable costs of production and transportation*, i.e., total profit.

By way of example, consider a company which produces a commodity on two plants and distributes it through three warehouses serving geographically scattered markets. Each plant has a given maximal capacity, and for each warehouse there is an upper limit to the amount that can be sold in the market it serves. The selling price differs between the markets, and unit costs of production between plants. The data are given in the following table.

		Unit transportation cost			Unit cost of production	Capacity
		Warehouse no.				
		1	2	3		
Factory no.	1	5	2	6	10	150
	2	2	6	8	12	200
Selling price per unit		18	17	19		
Maximum sales		100	90	95		

With slack variables — representing deliveries to a fictitious warehouse or shipments from a dummy factory — the transportation table becomes

x_{ij}		Warehouse no.				Total (capacity)
		1	2	3	4	
Factory no.	1	x_{11}	x_{12}	x_{13}	x_{14}	150
	2	x_{21}	x_{22}	x_{23}	x_{24}	200
	3	x_{31}	x_{32}	x_{33}	x_{34}	a
Total (maximum sales)		100	90	95	b	

However, the capacity of the dummy factory (a) and the maximum sales through the dummy warehouse (b) have not yet been fixed. In order not to preclude any feasible solution we must assign values to them such that the third factory equation and the fourth warehouse equation are still satisfied when the slack variables assume their largest possible values, i.e., when all structural variables are equal to zero. Then we have

$$x_{31} = 100, \ x_{32} = 90, \ x_{33} = 95$$

and

$$x_{14} = 150, \ x_{24} = 200,$$

which with $x_{34} = 0$ yields $a = 285$, $b = 350$. The table is now balanced.

Since total production is no longer fixed and, moreover, the selling price differs between markets (warehouses), total sales revenue now depends on the pattern of production and shipments. The objective function, therefore, must be total gross profit, so we have to calculate the unit profit coefficient associated with each activity, that is, selling price minus unit costs of production and transportation. Profit per unit of x_{11}, for example, is

$$18 - 10 - 5 = 3.$$

However, as the problem is to be formulated as a minimization problem, we set

$$c_{11} = -3,$$

and similarly for the other c_{ij} associated with the structural activities. The unit "cost" coefficients of the slack variables are, of course, set $= 0$.

The relevant information can then be compressed into the following table.

c_{ij}		Warehouse no.				Total (capacity)
		1	2	3	4	
Factory no.	1	−3	−5	−3	0	150
	2	−4	1	1	0	200
	3	0	0	0	0	285
Total (sales)		100	90	95	350	635

Note that some of the activities have positive c_{ij}, i.e., negative unit profit. However, such activities involving a loss will automatically be eliminated in the solution procedure. This follows from the way in which we fixed a and b such as to permit a basic solution in slack variables only[1].

E. Linear Investment Planning

1. In capital budgeting, or investment planning, the company is confronted with the problem of determining — within the limits of available financial resources — an *optimal investment plan*. Among the feasible alternatives — i.e., possible investment projects or combinations of projects — the one is selected which satisfies some criterion of optimality. We will assume that capital budgeting decisions are based on the criterion of maximum net present value, applying a rate of discount to the estimated future cash flows that characterize the alternatives[2].

Problems of this nature are traditionally solved by an *enumerative* procedure. Assume that there are two investment projects, I_1 and I_2, under consideration. A further possibility is the *zero investment* (I_0)

---

[1] *Exercise:* Formulate the three examples of unbalanced transportation problems as standard linear programming problems.

[2] As any textbook of capital budgeting indicates, there exist a number of other measures of investment worth, but the net present value criterion can be shown to rest on the firmest theoretical foundation.

which is characterized by zero cash flows; this possibility will have to be included because the best course of action may be to make no investments. We can then enumerate the logical possibilities, the alternative plans to choose from; in this case there are four logical possibilities as shown in the following table. We shall assume that they are all financially feasible.

Alternative (plan) no.	Project(s)	Net present value $Y=$
1	I_1 and I_2	Y_{1+2}
2	I_1	Y_1
3	I_2	Y_2
4	I_0	0

For each alternative the net present value (Y) of the cash flows is estimated; Y_1 is the net present value of project I_1 in case only this project is carried out, and similarly for Y_2, while Y_{1+2} refers to the case of both projects being carried out. The net present value of the zero investment is of course $=0$. The four plans are mutually exclusive and exhaust the set of possible plans. The optimal plan is then designated by the criterion
$$\max(Y_{1+2}, Y_1, Y_2, 0). \tag{8}$$

If I_1 and I_2 are mutually *independent*, i.e., if the cash flows associated with I_1 are the same whether or not I_2 is carried out and *vice versa*, we have
$$Y_{1+2} = Y_1 + Y_2$$
and (8) can be written
$$\max(Y_1 + Y_2, Y_1, Y_2, 0), \tag{9}$$
which can be translated into a very simple decision rule: Accept any project having a positive (or at least a non-negative) net present value. This rule applies to any number of independent projects, and we can do without the enumeration table; all we need to know is the signs of the $Y_j (j=1, 2, \ldots)$.

On the other hand, if the projects are *dependent* on each other so that the cash flows of one project are affected by the decision to accept or reject the other, this means that
$$Y_{1+2} \neq Y_1 + Y_2$$
so that (8) cannot be written in the form (9). In this case the selection of the optimal investment plan will have to start from a complete list of the feasible alternatives. However, with a large number of investment proposals many of which are dependent on one another, complete enumeration soon becomes impracticable. This is where linear program-

ming comes in as a practical method of handling large-scale investment planning problems[1] — provided that the dependencies are such that they can be described by linear restrictions.

2. The variables of a linear investment model are *decision variables* x_j associated with the respective projects I_j ($j=1, 2, \ldots$). They can assume the values 0 or 1 only, $x_j=0$ being interpreted as "reject project no. j" and $x_j=1$ as "accept". For example, $x_1=1$ and $x_2=0$ in the example above means that I_1 is carried out but I_2 is not (plan no. 2 in the table). The total net present value of a plan can then be expressed as a function of the x_j with the Y_j as coefficients, and the optimality criterion becomes

$$Y = Y_1 x_1 + Y_2 x_2 = \text{maximum} \qquad (10)$$

where

$$x_1, x_2 = 0 \text{ or } 1. \qquad (11)$$

If the preference function (10) is linear, i.e., if Y_1 and Y_2 are constants, and if the project dependencies (if any) can be described by linear relationships in x_1 and x_2, we have a linear programming problem, though a particular type in that the variables can assume no other values than 0 and 1[2].

3. If the projects I_1 and I_2 in the example above are mutually *independent*, Y_1 and Y_2 will be constants so that the preference function (10) is linear in x_1 and x_2. The four alternative plans are now expressed as the set of feasible solutions defined by (11), as shown in the following table. (10) is of course equivalent to (8) for $Y_{1+2} = Y_1 + Y_2$.

Alternative (plan) no.	$x_1 =$	$x_2 =$	$Y = Y_1 x_1 + Y_2 x_2$
1	1	1	$Y_1 + Y_2$
2	1	0	Y_1
3	0	1	Y_2
4	0	0	0

In this case the linear model is little more than a trivial reformulation of the traditional procedure.

If the projects are *dependent*[3], we can express the dependency relationship in terms of x_1 and x_2 as follows:

$$Y_1 = Y_1(x_2) \text{ and/or } Y_2 = Y_2(x_1). \qquad (12)$$

---

[1] Large-scale problems will of course have to be handled on a computer.

[2] *"0−1 programming"* is a special case of *integer programming*, i.e., linear programming where the variables can assume integral values only (0, 1, 2, ...). For solution procedures see Ch. VIII, E, below.

[3] For a discussion of dependencies see, e.g., H. Bierman and S. Smidt (1966), pp. 76ff.

For example, if project I_1 is a *complement* of I_2 — that is, the net cash flows expected from I_2 increase if I_1 is also undertaken — we have

$$Y_2(1) > Y_2(0)\text{[1]}. \qquad (13)$$

The reverse case,

$$Y_2(1) < Y_2(0) \qquad (14)$$

where acceptance of I_1 is harmful to I_2, indicates that I_1 is a *substitute* for I_2.

Dependency relations would appear at first sight to rule out a linear investment model, for inserting (12) in (10) we get

$$Y = Y_1(x_2) \cdot x_1 + Y_2(x_1) \cdot x_2 = \text{maximum},$$

that is, the preference function will no longer be linear: the coefficients change as we go from one feasible solution to another [2]. However, as we shall see, it is possible to describe the *extreme* — and practically important — cases of dependency by linear constraints in the decision variables, and in such a way as to preserve the linearity of the preference function as well.

4. The extreme case of the *complementarity* relation (13) is that in which no proceeds can be expected from project I_2 if I_1 is rejected; I_1 is a *prerequisite* of I_2. The net present value of I_2 will then be negative if I_1 is rejected (the cash flows expected from I_2 will consist of outlays only), i.e., $Y_2(0) < 0$. Hence I_2 will not be carried out if I_1 is rejected; in other words, plan no. 3 in the table above ($x_1 = 0$, $x_2 = 1$) is ruled out and instead of (13) we can write

$$x_2 \leq x_1 \qquad (15)$$

which together with the $0-1$ requirements (11) defines the set of feasible solutions as consisting of alternatives 1, 2, and 4.

Suppose by way of example that $Y_1 = 20$ whereas $Y_2 = 10$ or -50 according as to whether I_1 is accepted or rejected. Applying the traditional procedure we then have the four alternatives:

Plan no.	$Y=$
1	30
2	20
3	−50
4	0

No. 3 is obviously non-optimal and the best solution is plan no. 1.

---

[1] *Exercise:* Show that in this case we get $Y_{1+2} > Y_1 + Y_2$.
[2] *Exercise:* Show this in the Y column of the table above.

In a linear formulation, plan no. 3 is ruled out and we are left with three alternative solutions:

Plan no.	$x_1=$	$x_2=$	$Y = Y_1 \cdot x_1 + Y_2(x_1) \cdot x_2$
1	1	1	$20 \cdot 1 + 10 \cdot 1 = 30$
2	1	0	$20 \cdot 1 + 10 \cdot 0 = 20$
4	0	0	$20 \cdot 0 - 50 \cdot 0 = 0$

The optimal solution (plan no. 1) can now be determined by maximizing the linear function

$$Y = Y_1 \cdot x_1 + Y_2(1) \cdot x_2 = 20\,x_1 + 10\,x_2$$

subject to the linear restriction

$$x_2 \leq x_1 \qquad (15)$$

and the 0−1 requirements

$$x_1, x_2 = 0 \text{ or } 1. \qquad (11)$$

It is true that $Y_2 = -50$ instead of 10 in plan no. 4 (see the table), but obviously it makes no difference if $Y_2(1) = 10$ is used instead as the coefficient of x_2 since $Y=0$ anyway in plan no. 4. This ensures the linearity of the preference function.

Problems of this type can be solved by linear programming, using some method of solution especially designed for 0−1 programming. *Approximate* solutions can be determined by standard linear procedures (the simplex method) if the 0−1 requirements (11) are replaced by the restrictions

$$x_j \leq 1 \quad (j=1,2) \qquad (11\,\text{a})$$

$$x_j \geq 0 \quad (j=1,2), \qquad (11\,\text{b})$$

maximizing the preference function Y subject to side conditions of the types (15) and (11 a) and to the non-negativity requirements (11 b). If necessary, such a solution must subsequently be rounded off to a solution satisfying (11) and (15).

The extreme complementarity relation may be symmetrical so that the two projects are *mutual prerequisites*, This means that, in addition to (13), a similar relation holds for $Y_1(x_2)$ and that $Y_1(0)$ as well as $Y_2(0)$ will be negative. In terms of the x_j we now have not only $x_2 \leq x_1$ but also $x_1 \leq x_2$, or in other words

$$x_1 = x_2 \qquad (16)$$

which rules out plans 2 and 3. The optimal solution can now be determined by maximizing the linear preference function

$$Y = Y_1(1) \cdot x_1 + Y_2(1) \cdot x_2$$

subject to (16) and (11) or (11a—b).

The two cases of extreme complementarity, (15) and (16), can be illustrated geometrically as shown in Fig. 8. Replacing the 0−1 requirements (11) by (11a—b), the set of feasible solutions becomes the shaded triangle in Fig. 8a or the line segment connecting (0, 0) and (1, 1) in Fig. 8b.

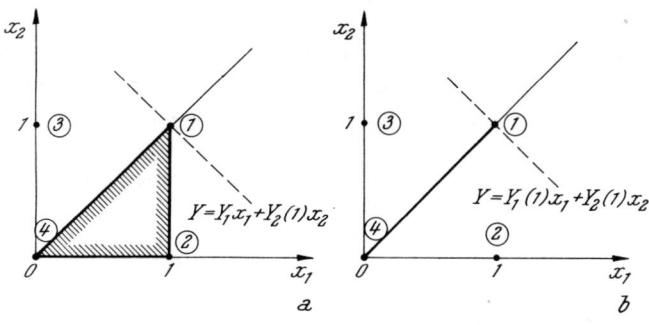

Figs. 8a and b

5. The opposite extreme is the case of strong *substitutability* where I_1 and I_2 are *mutually exclusive* either because it is physically impossible to carry out both projects (e. g., two factories would take up more land than there is available) or because $Y_1(1)$ and $Y_2(1)$ will both be negative (two factories producing the same commodity the demand for which is limited). In this case plan no. 1 can be ruled out and we can write

$$x_1 + x_2 \leq 1 \qquad (17)$$

which accommodates the remaining three plans. The optimal solution is determined by maximizing

$$Y = Y_1(0) \cdot x_1 + Y_2(0) \cdot x_2$$

subject to (17) and (11) or (11a—b).

For example, let Y_1 be $= -30$ or 20 according as to whether I_2 is undertaken or not, and similarly let Y_2 be $= -50$ or 10. The four alternatives are:

Plan no.	$Y =$
1	−80
2	20
3	10
4	0

In a linear formulation we have

Plan no.	$x_1=$	$x_2=$	$Y = Y_1(x_2) \cdot x_1 + Y_2(x_1) \cdot x_2$
2	1	0	$20 \cdot 1 - 50 \cdot 0 = 20$
3	0	1	$-30 \cdot 0 + 10 \cdot 1 = 10$
4	0	0	$20 \cdot 0 + 10 \cdot 0 = 0$

where the linear preference function

$$Y = Y_1(0) \cdot x_1 + Y_2(0) \cdot x_2 = 20 x_1 + 10 x_2$$

can be applied, maximizing it subject to (17) and (11); the negative coefficients (-50 in plan 2, -30 in plan 3) can safely be replaced by 10 and 20 respectively since they are multiplied by zeroes. The best plan will be no. 2.

A problem of this nature can be illustrated geometrically as in Fig. 9. The shaded triangle represents the set of feasible solutions when the $0-1$ requirements have been replaced by (11 a—b); otherwise only the corners of the triangle are feasible.

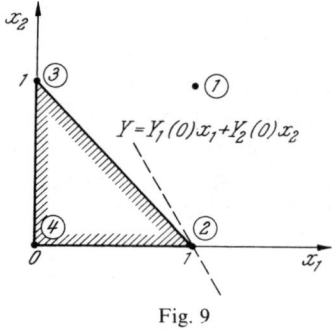

Fig. 9

If it is a foregone conclusion than one of the projects is to be carried out, (17) must be replaced by the equation

$$x_1 + x_2 = 1^1. \qquad (18)$$

6. The conclusion to be drawn from these examples is that a linear investment model in decision variables x_j can cope with investment projects which are *independent* as well as with projects which are either *mutually exclusive* or *prerequisites* of each other. (This holds for any number of projects.) All other dependency relationships — i.e., all but the extreme cases — violate the essential requirement of linearity and will have to be handled with traditional methods.

---

1 *Exercise:* Indicate the feasible solutions in Fig. 9 in this case.

Even in large-scale investment planning problems, solution by the traditional procedure is a fairly straightforward matter when the dependencies are of the simple forms (15)—(18). In practice, however, extreme dependencies — particularly strong substitutability — will frequently manifest themselves in a more or less disguised, but still linear form[1].

Suppose, for example, that two factory projects require 500 and 800 square metres of land respectively and that only 1,000 square metres are available. Such a *resource limitation* can be described by a linear inequality in the decision variables associated with the projects, x_1 and x_2 ($=0$ or 1):

$$500 x_1 + 800 x_2 \leq 1,000. \tag{19}$$

In this simple example the restriction might as well have been written in the more convenient form (17) since it is clear from (19) that the projects are mutually exclusive — $x_1 = x_2 = 1$ would violate (19)[2] — but in more complex planning problems with many projects and many dependencies a linear programming procedure will be computationally superior.

Similar restrictions may be due to other resource limitations. Let projects I_1 and I_2 require 3 and 2 specially skilled workers in the first year of operation; if only 4 men thus trained are available on the spot, we have

$$3 x_1 + 2 x_2 \leq 4. \tag{20}$$

Another example — important in practice — is financial restrictions caused by scarcity of funds for investment, or by the company's borrowing policy. Let the company's equity capital be $ 3,000,000, and assume that up to $ 9,000,000 can be borrowed in the capital market. There are two possible investment opportunities, the initial outlays of which are 10 and 5 million dollars. Then the choice is restricted by the inequality

$$10 x_1 + 5 x_2 \leq 12 \quad (=3+9). \tag{21}$$

There may be similar restrictions for the subsequent years of operation, expressing the requirement that available funds shall be non-negative at any time.

A different interpretation of the financial restriction (21) is the following. Assume that the company does not want its debt-equity ratio to exceed 3. With an equity capital of $ 3,000,000 available, this policy implies that at most 9 million should be borrowed.

In large-scale problems with many projects and many dependency relationships some of which are disguised, the traditional planning

---

[1] A comprehensive treatment of the various types of dependencies within a linear programming model of investment planning is given in I. Thygesen (1971), Ch. IV.

[2] *Exercise:* Illustrate geometrically that (19) and (17) impose the same restriction on the set of feasible solutions in the $0-1$ variables x_1 and x_2.

procedure tends to become impracticable because it is difficult or downright impossible to enumerate all the alternative plans that are logically possible and compatible with the dependencies. In such cases, linear programming (with $0-1$ variables) offers an operational method of investment planning.

7. The following simple example will illustrate how a linear investment model is set up.

A textile company is planning to build a spinning mill. Two alternative (mutually exclusive) projects are available, I_1 and I_2. In addition, the building of a cloth mill is considered, for which the existence of a spinning mill is a prerequisite; there are three alternative projects, I_3, I_4, and I_5. Estimates of net present values Y_j and initial outlays (building costs) A_j, in units of $ 1,000,000, are given in the following table.

$j=$	1	2	3	4	5
Y_j	6	4	3	5	6
A_j	16	12	8	12	16

At least 30 per cent of the total initial outlay (total building cost) is required to be covered by the company's equity capital, $ 8,400,000.

Translated into linear programming terms, the problem is the following:

$$Y = 6 x_1 + 4 x_2 + 3 x_3 + 5 x_4 + 6 x_5 = \text{maximum}$$

subject to

$$x_1 + x_2 \leq 1 \tag{22}$$

$$x_3 + x_4 + x_5 \leq 1 \tag{23}$$

$$x_3 + x_4 + x_5 \leq x_1 + x_2 \tag{24}$$

$$8.4 \geq 0.3 \left(\sum A_j x_j \right), \text{ i.e.}$$

$$16 x_1 + 12 x_2 + 8 x_3 + 12 x_4 + 16 x_5 \leq 28 \tag{25}$$

and

$$x_j = 0 \text{ or } 1 \quad (j = 1, 2, \ldots, 5) \tag{26}$$

or

$$x_j \leq 1, x_j \geq 0 \quad (j = 1, 2, \ldots, 5) \tag{26a}$$

where (24) expresses the restriction that a spinning mill is a prerequisite of the cloth mill[1].

---

[1] *Exercise:* Which of the side conditions are redundant?

V. Computational Procedures for Solving Linear Programming Problems

A. The Simplex Method

As we have seen in Ch. II[1], the simplex procedure can be described as a systematic method of examining the set of basic feasible solutions, starting in an arbitrary initial basis of m variables (activities) where m is the number of linear restrictions. If the initial basic solution does not satisfy the simplex criterion, we move to a neighbouring basis by replacing one of the basic variables, and so forth, until a basic feasible solution is attained in which all of the simplex coefficients are non-positive (in a minimization problem, non-negative). By the Fundamental Theorem, such a solution will be an optimal solution.

Suppose, for example, that we have the following *maximization* problem:

$$4 x_1 + 5 x_2 \leq 4{,}800$$
$$6 x_1 + 2 x_2 \leq 4{,}800 \tag{1}$$
$$x_1 \geq 0,\ x_2 \geq 0$$
$$13 x_1 + 10 x_2 = f = \text{maximum}.$$

Introducing slack variables x_3 and x_4, the problem can be rewritten as

$$\begin{aligned} 4 x_1 + 5 x_2 + 1 x_3 &= 4{,}800 \\ 6 x_1 + 2 x_2 + 1 x_4 &= 4{,}800 \\ 13 x_1 + 10 x_2 &= f = \text{max}. \end{aligned} \tag{2}$$

In problems of this type the slack variables form a convenient initial basis, since the solution for x_3, x_4, and f,

$$x_3 = 4{,}800 - 4 x_1 - 5 x_2 \tag{3a}$$
$$x_4 = 4{,}800 - 6 x_1 - 2 x_2 \tag{3b}$$
$$f = \phantom{4{,}80}0 + 13 x_1 + 10 x_2, \tag{3c}$$

---

1 See also Appendix, B.

can be found merely by rearranging terms in (2) without really having to "solve" the equations. For $x_1 = x_2 = 0$ we have the corresponding basic solution $x_3 = 4,800$, $x_4 = 4,800$, $f = 0$[1], which is seen to be feasible[2].

(3c) does not satisfy the simplex criterion; f is an increasing function of x_1 as well as of x_2 and the solution can be improved by making x_1 or x_2 positive instead of zero. We cannot introduce both of them at a time — the logic of the simplex procedure is such that only one basic variable is replaced when we shift to the next basis — so a choice has to be made. Either variable will do, but it usually pays *to introduce the variable whose simplex coefficient is largest*[3], in this case x_1. As x_1 increases (x_2 still $=0$), x_3 and x_4 become zero for $x_1 = 4,800/4 = 1,200$ and $x_1 = 4,800/6 = 800$ respectively (by (3a—b)), and the latter is the effective limit where x_4 drops out of the basis and is replaced by x_1.

The next step is to express the new basic variables, x_3 and x_1 (as well as f) in terms of x_2 and x_4. Again, this can be done without actually "solving" the original equations, for the solution can be found directly from (3) which carry precisely the same information as (1) or (2). In doing so we save a great deal of computational labour, particularly when the number of restrictions is large. The new equation for *the variable which comes into the basis* is found from *the old equation for the variable to be replaced* (i.e., from (3b)) by *dividing through by the coefficient of the new basic variable and rearranging terms*:

$$x_1 = \frac{4{,}800}{6} - \frac{2}{6} x_2 - \frac{1}{6} x_4 = 800 - \frac{1}{3} x_2 - \frac{1}{6} x_4. \qquad (4b)$$

The solution for *the variable that remains in the basis* is found by substituting this expression in (3a):

$$x_3 = 4{,}800 - 4 \cdot \left(800 - \frac{1}{3} x_2 - \frac{1}{6} x_4\right) - 5 x_2 = 1{,}600 - \frac{11}{3} x_2 + \frac{2}{3} x_4, \qquad (4a)$$

i.e., *the coefficients in the new equation* (including the constant term) *are those of the old equation plus the corresponding coefficients in* (4b) *multiplied by the coefficient of the new basic variable in the old equation*. A similar procedure applies to the preference function:

$$f = 0 + 13 \cdot \left(800 - \frac{1}{3} x_2 - \frac{1}{6} x_4\right) + 10 x_2 = 10{,}400 + \frac{17}{3} x_2 - \frac{13}{6} x_4. \qquad (4c)$$

---

1 *Exercise:* Assuming that (1) is a problem of optimal utilization of two machine capacities, what is the economic interpretation of this initial basic solution?

2 *Exercise:* Explain why a basis which gives a non-feasible solution cannot be used as a starting basis.

3 This is a purely empirical rule. Experience has shown that, if it is followed, the optimal solution will in most cases be reached in fewer steps than otherwise.

The coefficient of x_4 in (4c) is negative, hence the best thing we can do is to keep x_4 equal to zero; however, f is an increasing function of x_2, which is therefore the variable coming into the basis. The effective limit is $x_2 = 1,600 \cdot 3/11 = 436$, at which point x_3 becomes zero and drops out of the basis. Using the same substitution procedure as described above, the solution for the third set of basic variables is found to be

$$x_1 = 655 + \frac{1}{11} x_3 - \frac{5}{22} x_4 \qquad (5\text{b})$$

$$x_2 = 436 - \frac{3}{11} x_3 + \frac{2}{11} x_4 \qquad (5\text{a})$$

$$f = 12{,}875 - \frac{17}{11} x_3 - \frac{25}{22} x_4. \qquad (5\text{c})$$

This solution satisfies the simples criterion; the coefficients of the non-basic variables in the linear form are both negative so that f will have its largest value for $x_3 = x_4 = 0$. Thus the optimal solution is

$$x_1 = 655, \; x_2 = 436; \; f = 12{,}875^1.$$

B. The Simplex Tableau

1. This procedure, which led us step by step towards the optimal solution, can be expressed in a standard tabular form, the so-called *Simplex Tableau*. A simplex tableau is nothing but a table or matrix showing the coefficients of the problem at a particular step, and the coefficients in a tableau are computed from those of the previous tableau by an algorithm (a set of rules or formulas) which corresponds exactly to the substitution procedure described above by which we got from each basic solution to the next.

First, let us rearrange the original equations — i.e., the solution corresponding to the initial basis — as follows:

$$4{,}800 = 4 x_1 + 5 x_2 + 1 x_3 + 0 x_4$$
$$4{,}800 = 6 x_1 + 2 x_2 + 0 x_3 + 1 x_4$$
$$0 = -13 x_1 - 10 x_2 + 0 x_3 + 0 x_4 + f,$$

which is merely another way of writing (2) or (3). The solutions for the second and third sets of basic variables can be rewritten in a similar fashion, with the constant terms — i.e., the basic solution — to the left of the equality sign and all the variables on the right-hand sides. The order

---

[1] *Exercise:* Draw the geometric picture of the problem (1) and indicate the three extreme points (corners of the area of feasible solutions) which correspond to the three computational steps (3), (4), and (5).

in which the variables are entered is of course arbitrary, only the same order has to be preserved through all of the steps. Similarly, the order of the equations should be preserved, in the sense that the solution for a variable which comes into the basis takes the place of the equation for the variable that goes out.

A simplex tableau is a table displaying the coefficients of one of these equation systems; to each step (basis) corresponds a simplex tableau, and the optimal solution to the present problem can be read off the final tableau. The simplex tableaux corresponding to the three steps in which the present problem is solved are as follows:

				0	13	10	0	0	
				P_0	P_1	P_2	P_3	P_4	
I	0	P_3		4,800	↓ 4	5	1	0	1,200
	0	P_4		←4,800	6	2	0	1	800
	$z_j - c_j$			0	−13	−10	0	0	
II	0	P_3		←1,600	0	↓ $\frac{11}{3}$	1	$-\frac{2}{3}$	436
	13	P_1		800	1	$\frac{1}{3}$	0	$\frac{1}{6}$	2,400
	$z_j - c_j$			10,400	0	$-\frac{17}{3}$	0	$\frac{13}{6}$	
III	10	P_2		436	0	1	$\frac{3}{11}$	$-\frac{2}{11}$	
	13	P_1		655	1	0	$-\frac{1}{11}$	$\frac{5}{22}$	
	$z_j - c_j$			12,875	0	0	$\frac{17}{11}$	$\frac{25}{22}$	

The order of the variables or activities is indicated in the head of the table, P_j being the column of coefficients associated with x_j; the constant terms — i.e., the basic solutions — are entered in the column denoted by P_0[1]. The coefficients of the variables in the original preference function, the c_j, are listed in the head of the table over the P's. The stub of the table indicates the bases and the c's associated with the basic variables. The $z_j - c_j$ row in each tableau gives the simplex coefficients and — in the P_0 column — the value of the preference function in the basic solution.

The procedure for calculating the coefficients of Tableau II from those of Tableau I corresponds exactly to the procedure by which we transformed

---

[1] Alternatively, the columns may be denoted by $x_0, x_1, x_2, \ldots$.

(3) into (4), except that *the variable to be introduced* into the basis is now designated by the *most negative* coefficient in the $z_j - c_j$ row of Tableau I, in this case the coefficient of x_1 (column P_1). *The basic variable to be replaced* is determined by calculating the upper limits to the new basic variable, i.e., by dividing the elements of the P_0 column by the corresponding elements in column P_1; these limits are listed in the margin to the right of the tableau. (The division is not carried out for negative or zero elements in the P_1 column[1].) The lowest of these limits is here the one in the P_4 row, hence x_4 is to be replaced by x_1 in the basis, as indicated by the arrows. After indicating the new basis by writing P_3 and P_1 in the stub, we are now ready to fill in Tableau II.

We first calculate *the row of the new basic variable*, i.e., the P_1 row. This is done by *dividing the row of the old basic variable by the coefficient of the new basic variable*, 6. (Cf. the transformation of (3 b) into (4 b) above.) This coefficient, printed in bold-faced type, is located at the intersection of the new basic variable's column and the old variable's row in Tableau I.

Next we proceed to calculate *the row of the variable which remains in the basis*, the row labelled P_3. In accordance with the procedure used above for transforming (3a) into (4a), *each element in the new P_3 row* is computed as *the coefficient in the old P_3 row minus the corresponding coefficient in the new P_1 row* (which we have just calculated) *multiplied by the coefficient appearing at the intersection of the P_1 column and the P_3 row in the old tableau* (in this case, 4). In the general case of more than one variable remaining in the basis this procedure applies to each of the corresponding rows. Moreover, *the same procedure is used to transform the $z_j - c_j$ row*.

The following table shows how this procedure works in transforming Tableau I into Tableau II.

			0 P_0	13 P_1	10 P_2	0 P_3	0 P_4
	0	P_3	$4{,}800 - 4 \cdot 800$ $= 1{,}600$	$4 - 4 \cdot 1$ $= 0$	$5 - 4 \cdot \frac{1}{3}$ $= \frac{11}{3}$	$1 - 4 \cdot 0$ $= 1$	$0 - 4 \cdot \frac{1}{6}$ $= \frac{-2}{3}$
II	13	P_1	$4{,}800 : 6$ $= 800$	$6 : 6$ $= 1$	$2 : 6$ $= \frac{1}{3}$	$0 : 6$ $= 0$	$1 : 6$ $= \frac{1}{6}$
	$z_j - c_j$		$0 + 13 \cdot 800$ $= 10{,}400$	$-13 + 13 \cdot 1$ $= 0$	$-10 + 13 \cdot \frac{1}{3}$ $= \frac{-17}{3}$	$0 + 13 \cdot 0$ $= 0$	$0 + 13 \cdot \frac{1}{6}$ $= \frac{13}{6}$

---

[1] *Exercise:* Explain why.

In a similar fashion, Tableau II is transformed. The resulting third tableau has *no negative coefficients in the* z_j-c_j *row*, i.e., the simplex criterion is satisfied and the third basic solution,

$$x_1 = 655, \quad x_2 = 436, \quad f = 12{,}875,$$

which appears in the P_0 column of Tableau III, is an optimal solution[1].

These rules for transforming the simplex tableau can of course be expressed algebraically in a set of formulas, the "*Simplex Algorithm*". The formulas are derived in the Appendix[2].

A number of computational short cuts follow directly from the logic of the procedure (or from the formulas):

(i) It is not necessary to calculate the elements of the basic columns in the new tableau. In the column of a basic variable x_i, the coefficient appearing at the intersection of the P_i column and the P_i row will always be 1 and all other coefficients in the column will be zero, including $z_i - c_i$. (Cf. columns P_1 and P_3 in the second tableau, and P_1 and P_2 in the third.)

(ii) Whenever a coefficient in the *column* of the variable to be *introduced* into the basis is zero, the corresponding *row* can be transferred unchanged to the new tableau.

(iii) When a coefficient in the *row* of the variable to be *replaced* is zero, the corresponding *column* is transferred unchanged[3].

In the example above, the coefficients in the z_j-c_j row were calculated by use of the simplex algorithm. However, the result can be checked by computing the simplex coefficients in a different way. In the first tableau, for example, the coefficient $z_1 - c_1 = -13$ indicates that a unit increase in the value of x_1 (starting from the basic solution in which x_1 is zero) will increase the value of the preference function by 13 units. Now the first two rows in the tableau — or the equivalent equations (3a—b) — show that if x_1 is increased by 1, the values of x_3 and x_4 will be reduced by 4 and 6 units respectively. The effects of these changes on the value of f can be computed by multiplying by the respective coefficients in the original preference function, the c_j. Thus the "direct effect" of the increase in x_1 is $c_1 \cdot 1 = c_1 = 13$, whereas the "indirect effect" — the decrease in f brought about by the decrease in x_3 and x_4 — is

$$z_1 = c_3 \cdot 4 + c_4 \cdot 6 = 0 \cdot 4 + 0 \cdot 6,$$

which happens to be zero because both of the c_j coefficients are zero. The total, or "net" effect per unit of increase in x_1 is

---

1 *Exercise:* Why cannot the simplex calculations be confined to the P_0 column (i.e., the basic feasible solutions) and the z_j-c_j row (the simplex coefficients), which contain all the information we are really interested in?

2 See Appendix, C.

3 *Exercise:* Explain why (i)–(iii) must always hold.

$$c_1 - z_1 = c_1 - (c_3 \cdot 4 + c_4 \cdot 6) = 13 - 0 = 13,$$

which is the coefficient of x_1 in (3c), as it should be. The corresponding coefficient in the first simplex tableau is

$$z_1 - c_1 = -13.$$

In other words, the simplex coefficient in the P_1 column can be calculated by multiplying the elements in the P_3 and P_4 rows (the rows of the basic variables) by the corresponding c's, adding the products to get z_1, and substracting c_1. It is easily verified that similar operations in all other columns of the three tableaux tally with the simplex coefficients already present in the $z_j - c_j$ rows[1]. Thus, in general, the simplex coefficients which have been calculated by the simplex algorithm can alternatively be computed as

$$z_j - c_j = \sum_i c_i x_{ij} - c_j \qquad (6)$$

where c_i is the coefficient of the basic variable x_i in the original preference function and x_{ij} is the element appearing at the intersection of the P_i row and the P_j column[2,3]. The c_i are listed in the stub and the c_j in the head of the table. This result holds for all j, including the columns of the basic variables (where $z_j - c_j$ will always be $=0$), as well as for the P_0 column if for convenience we write $c_0 = 0$ at the top of the column.

2. The simplex procedure by which we have just solved the problem (1) is applicable to all maximization problems involving inequalities of the type $\leq$ in the structural variables. Problems of *minimizing the preference function subject to inequalities of the type* $\geq$ are solved by a slightly modified procedure.

In the first place, the slack variables can no longer be used as a starting basis because their negative signs will make the corresponding basic solution negative (it being assumed that the right-hand sides of the inequalities are positive[4]). Any other set of m variables (some of which may be slack variables) can be used, provided that they yield a non-negative solution. However, it is usually more convenient to introduce, in addition to the slack variables, a set of m "*artificial variables*" to form an initial basis, particularly when m is large. The ith of these new variables is given the

---

[1] *Exercise:* Carry out these calculations.

[2] The validity of this result can easily be checked using the formulas of the simplex algorithm, cf. Appendix, C.

[3] *Exercise:* Assuming that (1) is a problem of optimal utilization of machine capacities, indicate the economic interpretation of the simplex coefficient as computed by the formula (6).

[4] *Exercise:* Can the slack variables be used as an initial basis if the right-hand sides are all negative?

coefficient 1 in the ith equation and zero coefficients in all other equations, so that the first basic solution will be the constant terms of the restrictions. In order to ensure that these artificial slack variables will appear with zero values in the final solution, as they should since they have no meaningful interpretation in the concrete problem to be solved, each of them is given a very large positive coefficient M in the preference function. Then a basic solution which has artificial variables in the basis can never be an optimal solution, because the corresponding value of f — the function to be minimized — can be made as large as desired by making M sufficiently large[1].

A similar device can be used to provide a non-negative initial solution if some or all of the restrictions have the form of *equations* in the structural variables, or if the problem has *inequalities of both types* ($\leq$ and $\geq$); in the latter case there will be slack as well as artificial variables in the initial basis. (If it is a *maximization* problem, the artificial variables should have a large *negative* coefficient $-M$ in the linear preference function.)

In the second place, the variable to be *introduced* at each step is designated by the *largest positive coefficient* in the $z_j - c_j$ row of the simplex tableau; and the *optimal* programme is characterized by there being *no* positive $z_j - c_j$ coefficients in the tableau.

The following example will illustrate the procedure. Suppose that we have the problem (12) of Ch. II[2], which may be interpreted, for example, as a blending problem of the "diet" type. The two inequalities are transformed into equations by introducing slack variables x_4 and x_5. However, the basic solution for the slack variables will be negative and thus nonfeasible:

$$x_4 = -5, \ x_5 = -4.$$

A positive initial solution may be obtained by choosing a "mixed" basis, for example (x_1, x_5). However, another feasible starting basis may be provided by two artificial variables x_6 and x_7; the problem then becomes

$$\begin{aligned} 2x_1 + 4x_3 - 1x_4 + 1x_6 &= 5 \\ 2x_1 + 3x_2 + 1x_3 - 1x_5 + 1x_7 &= 4 \\ x_1 \geq 0, \ x_2 \geq 0, \ x_3 \geq 0, \ x_4 \geq 0, \ x_5 \geq 0, \ x_6 \geq 0, \ x_7 &\geq 0 \\ 4x_1 + 2x_2 + 3x_3 + 0x_4 + 0x_5 + Mx_6 + Mx_7 &= g = \min., \end{aligned} \quad (7)$$

where the unspecified M can be interpreted as a huge penalty for underfulfilment of the requirements (or as a prohibitive price associated with the use of two fictitious ingredients).

---

[1] It is not necessary to specify the numerical value of M in the computations, although in some cases it is convenient to set it equal to some very large number (when the problem is solved on a computer).
[2] See p. 10.

The Simplex Tableau

			P_0	4 P_1	2 P_2	3 P_3	0 P_4	0 P_5	M P_6	M P_7	
I	M	P_6	5	2	0	4	-1	0	1	0	$5/4$
	M	P_7	4	2	3	1	0	-1	0	1	4
		z_j	$9M$	$4M$	$3M$	$5M$	$-M$	$-M$	M	M	
		z_j-c_j	$9M$	$4M-4$	$3M-2$	$5M-3$	$-M$	$-M$	0	0	
II	3	P_3	$\frac{5}{4}$	$\frac{1}{2}$	0	1	$\frac{-1}{4}$	0	$\frac{1}{4}$	0	
	M	P_7	$\frac{11}{4}$	$\frac{3}{2}$	3	0	$\frac{1}{4}$	-1	$\frac{-1}{4}$	1	$\frac{11}{12}$
		z_j	$\frac{15}{4}+M\cdot\frac{11}{4}$	$\frac{3}{2}+M\cdot\frac{3}{2}$	$3M$	3	$\frac{-3}{4}+M\cdot\frac{1}{4}$	$-M$	$\frac{3}{4}-M\cdot\frac{1}{4}$	M	
		z_j-c_j	$\frac{15}{4}+M\cdot\frac{11}{4}$	$\frac{-5}{2}+M\cdot\frac{3}{2}$	$-2+M\cdot 3$	0	$\frac{-3}{4}+M\cdot\frac{1}{4}$	$-M$	$\frac{3}{4}-M\cdot\frac{5}{4}$	0	
III	3	P_3	$\frac{5}{4}$	$\frac{1}{2}$	0	1	$\frac{-1}{4}$	0	$\frac{1}{4}$	$\frac{-1}{3}$	
	2	P_2	$\frac{11}{12}$	$\frac{1}{2}$	1	0	$\frac{1}{12}$	$\frac{-1}{3}$	$\frac{-1}{12}$	$\frac{1}{3}$	
		z_j	$\frac{67}{12}$	$\frac{5}{2}$	2	3	$\frac{-7}{12}$	$\frac{-2}{3}$	$\frac{7}{12}$	$\frac{2}{3}$	
		z_j-c_j	$\frac{67}{12}$	$\frac{-3}{2}$	0	0	$\frac{-7}{12}$	$\frac{-2}{3}$	$\frac{7}{12}-M$	$\frac{2}{3}-M$	

With (x_6, x_7) as a starting basis, the coefficients in the P_6 and P_7 rows of the first simplex tableau are those of the two equations. The simplex coefficients, however, can *not* be taken directly from (7) since we must first express g as a function of the non-basic variables only, by substituting the solution for x_6 and x_7 in the g function. (Alternately the $z_j - c_j$ of the first tableau may be calculated by the column operations described above, i.e., by computing z_j and subtracting c_j.) As the table above shows, M will then appear in the simplex coefficients of the first tableau. For large values of M, the largest (the "most positive") $z_j - c_j$ will be $z_3 - c_3$, hence x_3 is the new basic variable. Following otherwise the usual simplex procedure, the optimal solution appears in the third tableau, where none of the $z_j - c_j$ are positive and the artificial variables have been thrown out of the basis. The computations are shown in the table. The last row of the third tableau reads

$$g = \frac{67}{12} + \left(M - \frac{7}{12}\right) x_6 + \left(M - \frac{2}{3}\right) x_7 + \frac{3}{2} x_1 + \frac{7}{12} x_4 + \frac{2}{3} x_5;$$

since the coefficients of the five non-basic variables are all positive for sufficiently large values of M, g will have a minimum for all of these variables $= 0$, i.e., the corresponding basic feasible solution

$$x_3 = \frac{5}{4}, \; x_2 = \frac{11}{12}; \; g = \frac{67}{12}$$

is an optimal solution[1].

C. Alternate Optima and Second-Best Solutions

1. As we have seen, an optimal solution to a maximization problem is characterized by all of the $z_j - c_j$ being *non-negative*[2] (except of course for the simplex coefficients of the basic variables, which are always zero). If they are all strictly *positive*, the optimal basic solution is *unique*, because obviously a positive value to any non-basic variable will reduce the value of the preference function. However, if there are *zeroes* among the simplex coefficients of the non-basic variables in the final tableau, this will indicate the existence of *alternate optimal basic solutions*, for a non-basic variable whose $z_j - c_j$ is zero may be introduced into the basis without affecting the (maximum) value of f[3]. Thus the alternate solution can be computed

---

1 A number of variants of the simplex method and alternative computational techniques for computer solution of large-scale or specialized linear problems have been developed in recent years.

2 In minimization problems, *non-positive*.

3 Alternate optima may also occur without zero coefficients appearing in the $z_j - c_j$ row, namely, when the optimal basic solution is degenerate so that one or more basic variables are zero. These variables can be replaced in the basis by others without affecting the value of the preference function. However, such "alternate" optima are in effect identical. Cf. Ch. V, E, below.

Alternate Optima and Second-Best Solutions

by carrying the calculations one step further, i.e., by proceeding in the usual fashion to a new tableau where the variable in question has come into the basis. Repeating this procedure for all other variables whose simplex coefficients were zero, we will get the complete set of basic optimal solutions. Moreover, any "convex combination" (weighted average) of these alternate optimal solutions will be an optimal — though nonbasic — solution; thus there exist an infinite number of optimal solutions, but they can all be "generated" by the basic optimal solutions, which are finite in number.

In practical applications of linear programming, alternate optimal solutions, if any, should always be calculated; although they are all equally good solutions to the mathematical optimization problem as such, one may well be better than the others for "outside" reasons (e.g., because it involves the least setup time).

2. Similar considerations may call for the computation of second-best, third-best solutions, etc.

The second-best basic solution is to be found among the "neighbouring" basic solutions, i.e., it will be only one step removed from the optimal programme[1,2]. Thus the proper procedure for finding it is to introduce the non-basic variables into the basis in turn, each time starting from the tableau in which the optimal solution is displayed. The second-best solution will evidently be designated by *the variable for which the product of the simplex coefficient $z_j - c_j$ and the effective limit is smallest* (but $\neq 0$), since this is the amount by which the value of the preference function is affected by the change of basis.

As an example, let us go back to the beer blending problem of Ch. IV, A, where — as will be remembered — the second-best solution claims considerable attention because of possible differences of taste quality. The simplex tableau corresponding to the optimal solution (6) is displayed as Tableau I below.

There are three non-basic variables, x_4, x_5, and x_{10}. If x_4 is made positive, the *net increase* in total cost of ingredients will be 0.82 crowns *per unit of* x_4, as indicated by the simplex coefficient $z_4 - c_4$. The effective limit to the increase in x_4 is found to be $50.78:4.05 = 12.54$, at which point x_4 replaces x_8 in the basis. Consequently the cost of making this new blend will be $0.82 \cdot 12.54 = 10.28$ crowns higher than the cost of the optimal blend. Similarly, the introduction of x_5 or x_{10} into the basis — again

---

1 *Exercise:* Explain why this must be so, and give a geometrical illustration of a case involving only two structural variables.

2 Of course the second-best solution here means the second-best *basic* solution. Obviously the best and the second-best solution can be combined to give intermediate solutions in more than m variables.

			0	44	50	64	90	0	0	0	0	0	0
		P_0	P_1	P_2	P_3	P_4	P_5	P_6	P_7	P_8	P_9	P_{10}	P_{11}
I	0 P_1	9.84	1			−0.29	−0.54					0.02	
	44 P_2	19.67		1		0.22	−0.08					−0.06	
	50 P_3	70.49			1	1.07	1.62					0.04	
	0 P_6	0.22				0.01	0.01	1				0.00	
	0 P_7	0.38				−0.01	−0.01		1			0.00	
	0 P_8	50.78				4.05	6.70			1		−0.30	
	0 P_9	149.22				−4.05	−6.70				1	0.30	
	0 P_{11}	500										1.00	1
	$z_j - c_j$	4,389.98				−0.82	−12.52					−0.64	
II a	0 P_1	13.48											
	44 P_2	16.91											
	50 P_3	57.07											
	0 P_6	0.09											
	0 P_7	0.51											
	64 P_4	12.54											
	0 P_9	200.00											
	0 P_{11}	500.00											
	$z_j - c_j$	4,400.26											
II b	0 P_1	13.93											
	44 P_2	20.28											
	50 P_3	58.21											
	0 P_6	0.14											
	0 P_7	0.46											
	90 P_5	7.58											
	0 P_9	200.00											
	0 P_{11}	500.00											
	$z_j - c_j$	4,484.88											
II c	0 P_{10}	492.00											
	44 P_2	49.19											
	50 P_3	50.81											
	0 P_6	0.22											
	0 P_7	0.38											
	0 P_8	198.38											
	0 P_9	1.62											
	0 P_{11}	8.00											
	$z_j - c_j$	4,704.86											

starting from Tableau I — will increase total cost of ingredients by 12.52 · 7.58 = 94.90 and 0.64 · 492.00 = 314.88 crowns respectively. Thus the second-best solution is to introduce x_4, i.e. to add the fourth ingredient at the expense of x_2 and x_3 (at the same time, however, adding more water to the mix since x_1 is an increasing function of x_4) so far as this is compatible with the lower limit specified for the colour index of the mixture. The solutions corresponding to the three substitutions are shown in Tableau IIa—c, of which IIa represents the second-best solution[1,2].

D. Computational Short Cuts

As we have seen, the simplex procedure for solving linear programming problems is extremely simple and straightforward. No mathematical operations in the proper sense of the word are involved, nor do the numerical calculations go beyond the four basic arithmetical operations. The routine can easily be picked up by anyone. The only complicating factor is the vast number of operations to be performed in solving problems of large dimensions; in many such cases solution by means of even an electric desk computer may be impracticable and the problem will have to be solved by an electronic computer.

The amount of computational labour involved in solving a programming problem is determined by (i) the number of steps or iterations (simplex tableaux) and (ii) the number of operations to be performed per iteration.

The *number of steps* will of course depend on the number of restrictions, m; practical experience has shown that it will usually be between m and $2m$ when slack or artificial variables are used as a starting basis. On a superficial view it might perhaps have been expected that it would take at most $m+1$ iterations[3] to get to the optimal solution, since the initial basis consists of m variables of which at most m will have to be replaced. This would be true if the simplex procedure could make no "detours"; however, it happens quite frequently during the computations that a variable is introduced into the basis only to be thrown out later[4], or that a basic variable which has been replaced returns to the basis at some later stage.

In some cases the amount of computational work can be considerably reduced by starting from a basis which is known, or at least supposed, to be closer to the optimal basis than the usual initial basis of slack or

---

1 *Exercise:* Fill in the blank spaces in the three tableaux.

2 *Exercise:* Does it follow from the reasoning above that Tableaux IIb and IIc represent the third-best and the fourth-best solutions?

3 Including the initial tableau.

4 This happens twice during the solution of the ice-cream problem of Ch. III. All of the four artificial variables that constitute the initial basis are replaced, which requires 5 tableaux (i.e., 4 simplex tableaux in addition to the initial one), but the 2 detours require 2 additional iterations so that the problem is solved in 7 steps.

artificial variables. For example, the company planner will often have a preconceived notion — based on practical experience, intuition, or the like — of the location of the optimal solution, and wants to have it tested for optimality; or inspection of the structure of the concrete problem (the matrix of coefficients) may suggest that such and such variables will have to be in the optimal basis. If the choice of a starting basis is guided by such considerations, the number of computational steps required for solving the problem will be reduced, in some cases to a single step. Against this must be weighed the extra work involved in having to solve the equations for this particular set of basic variables. If the structure of the problem is such as to permit easy solution for these variables — i.e., if there are plenty of zeroes among the coefficients and if m is not too large — the procedure may save a great deal of computational work[1].

The *number of arithmetical operations* to be performed at each step will depend upon the number of coefficients involved, i.e., on the dimension of the problem, $m \cdot n$, where m is the number of linear restrictions and n is the number of structural variables. However, as we have already seen, some of the operations become trivial if there are zeroes among the coefficients, as will usually be the case in practical applications; zeroes in the column of the variable to be introduced or in the row of the variable to be replaced make it possible to carry over entire rows or columns to the next tableau.

A number of computational short cuts have been developed for special types of linear programming problems, most of them exploiting particular patterns of zeroes in the coefficient matrix. One example is the simplified version of the simplex method which is used to solve problems of the *transportation* type, where the coefficients are ones and zeroes forming a regular pattern; this procedure will be dealt with in the last section of the present chapter.

E. The Case of Degeneracy

In well-behaved linear programming problems, the basic solutions corresponding to the successive steps of the simplex procedure will all be positive. There are cases, however, in which one or more of the m basic

---

[1] The ice-cream problem of Ch. III is a case in point. By inspection of the coefficients, the region within which the optimal solution was to be found could be narrowed down considerably, and the particular basic solution which was actually used by the company was known (and was within the region). The problem was easily reduced such as to involve only four equations, and the majority of the coefficients were zero. The first basic solution turned out to be optimal, whereas it would have taken seven iterations (simplex tableaux) to solve the problem starting from an initial basis of artifical variables.

variables becomes zero at some stage of the calculations so that zeroes appear in the P_0 column of the simplex tableau. The basic solution is then said to be *degenerate*[1,2].

Degeneracy may occur at the very first step; when there are zeroes among the constant terms in the equations (i.e., the right-hand terms of the linear restrictions in their original form), the corresponding slack or artificial variables in the initial basis will be zero. Or a degenerate solution may appear at a later stage because two or more basic variables set the same effective limit to the new basic variable so that the next tableau will have fewer than m positive variables.

Suppose, for example, that we have the maximum problem shown below.

				$\frac{7}{4}$	3	4	0	0	
			P_0	P_1	P_2	P_3	P_4	P_5	
I	0	P_4	1	2	1	1	1	0	1
	0	P_5	2	1	2	4	0	1	$\frac{1}{2}$
	$z_j - c_j$		0	$\frac{-7}{4}$	-3	-4	0	0	
II	0	P_4	$\frac{1}{2}$	$\frac{7}{4}$	$\frac{1}{2}$	0	1	$\frac{-1}{4}$	1	$\frac{7}{2}$
	4	P_3	$\frac{1}{2}$	$\frac{1}{4}$	$\frac{1}{2}$	1	0	$\frac{1}{4}$	1	$\frac{1}{2}$
	$z_j - c_j$		2	$\frac{-3}{4}$	-1	0	0	0	
III	0	P_4	0	$\frac{3}{2}$	0	-1	1	$\frac{-1}{2}$	0
	3	P_2	1	$\frac{1}{2}$	1	2	0	$\frac{1}{2}$	2
	$z_j - c_j$		3	$\frac{-1}{4}$	0	2	0	$\frac{3}{2}$	

---

[1] A degenerate solution in k positive variables implies that a system of m linear equations in k variables ($m > k$) has a solution; clearly this is an exceptional case. The attention given to the problem of degeneracy in the literature is quite out of proportion to the actual frequency of degenerate solutions in practical problems, as well as to the relatively insignificant computational difficulties caused by degeneracy. Except for the rather special transportation problem, degenerate solutions do not occur very frequently in practice. (However, two of the concrete problems in Ch. IV had degenerate basic solutions.)

[2] *Exercise:* Construct a geometric example (in two structural variables) in which one of the basic feasible solutions is degenerate.

No difficulties are met with in transforming the first tableau. In Tableau II, however, the two limits listed in the margin coincide so that both x_3 and x_4 become zero when x_2, the variable to be introduced, is increased as far as compatible with the non-negativity requirements. Thus there will be only one positive variable in the third basic solution. However, in order to be able to continue our exploration of the basic solutions, we must retain one of the zero variables in the basis, entering a zero in the P_0 column of the third tableau. Either one will do; let us, for example, consider x_3 as the variable to be replaced by x_2 as a basic variable and x_4 as the (zero) variable to remain in the basis, as shown in Tableau III.

The simplex criterion is not satisfied by the third solution, so we proceed to a fourth tableau. Introducing x_1 into the basis, x_4 becomes the variable to be replaced. Since the effective limit to x_1 is zero $(=0:3/2)$, the substitution is of a purely formal nature, so far as the basic solution is concerned; x_1 is simply entered as a zero variable in the new basic solution instead of x_4, so that we get a new degenerate basic solution as shown in Tableau IV.

	7/4	P_1	0	1	0	$\dfrac{-2}{3}$	$\dfrac{2}{3}$	$\dfrac{-1}{3}$
IV	3	P_2	1	0	1	$\dfrac{7}{3}$	$\dfrac{-1}{3}$	$\dfrac{2}{3}$
	z_j-c_j		3	0	0	$\dfrac{11}{6}$	$\dfrac{1}{6}$	$\dfrac{17}{12}$

Here the simplex criterion is satisfied, i.e., the fourth basic solution is an optimal solution[1].

The reader will have perceived that the basic solutions of Tableaux III and IV are *identical*; in either case we have the degenerate solution in one positive variable[2]

$$x_2 = 1;\ f = 3.$$

All other variables are equal to zero, and it makes no real difference to the solution whether one of them — no matter which one — is formally considered as a basic variable having the value 0. It follows that the basic solution displayed in Tableau III was an *optimal* solution although the simplex criterion was *not* satisfied. Thus the example illustrates the fact that the simplex criterion is not a *necessary* condition for a basic

---

1 *Exercise:* Solve the problem with x_4, instead of x_3, as the variable to be replaced in the basis of Tableau II.

2 The existence of a degenerate solution was obvious from the outset; since P_0 is proportional to P_2 in the first tableau, the restrictions will be satisfied by a positive solution in x_2 only.

solution to be optimal when the solution is degenerate[1]. However, this fact caused no computational difficulties; by proceeding to Tableau IV we arrived at a new basis where all of the simplex coefficients turned out to be positive, and which is known with certainty to be optimal since the simplex criterion *always* provides a *sufficient* condition of optimality.

Thus degeneracy would not appear to present any computational difficulties; the fact that a degenerate solution occurred at a certain step did not prevent us from carrying on with the usual simplex procedure. However, in the case of degeneracy there is always a theoretical risk of the procedure becoming repetitive before an optimal solution has been attained, a phenomenon known as "cycling". When the iterative procedure passes through a series of basic solutions which are identical except for the alternating zero variables that are formally treated as belonging to the basis (cf. Tableaux III and IV above), it may happen that one of these variables, after having been thrown out at an earlier stage, returns to the basis before we are through with the series. This means that we are back at an earlier basis so that the procedure will repeat itself *ad infinitum* instead of converging in a finite number of steps to a solution for which the simplex criterion is satisfied[2].

Although no practical examples of cycling have ever been observed — indeed, it is difficult even to construct a hypothetical example that leads to repetitive cycles of basic solutions[3] — the possibility of cycling does exist. To overcome this theoretical difficulty, a rule of thumb has been devised which effectively precludes the occurrence of cycling. As we have seen, degeneracy occurs when the effective limit to the new basic variable is set jointly by several variables, i. e., when the variable to be removed from the basis is not uniquely determined. When such a "tie" occurs between two or more basic variables — such as between x_3 and x_4 in Tableau II above — the ratios x_{i1}/x_{ik} (k being the number of the new basic variable) are computed for each of the corresponding rows, i.e., the ratio of the coefficient in the P_1 column to that occurring in the column of the variable to be introduced; the smallest[4] of these ratios designates the variable to be replaced. Thus in Tableau II we calculate

---

1 Cf. Ch. II. p. 14. and Appendix. B. The point is that, even though the sign of the simplex coefficient $z_1 - c_1$ in Tableau III indicates that the solution could be improved if x_1 could be made positive, this is not feasible because x_4, which is zero in the basic solution, would then become negative.

2 *Exercise:* Why can cycling occur only in the case of degeneracy?

3 One of the very few examples that exist is constructed by Beale, cf. E. M. L. Beale (1955) or E. O. Heady and W. Candler (1958), pp. 146ff. Another example is found in N. Nielsen (1956).

4 By "smallest" is meant *algebraically* smallest; thus, for example, -2 is smaller than 1, and -3 is smaller than -2.

the ratios (7/4):(1/2)=7/2 and (1/4):(1/2)=1/2 and list them to the right in the margin; the latter is smaller and, since it occurs in the P_3 row, x_3 is the variable to be replaced. If the tie had persisted, i.e., if the ratios had been equal, the decision would have been determined by comparing the ratios x_{i2}/x_{ik}, and so forth. Since no two rows can ever be identical[1], the tie must eventually be broken; thus the choice of the variable to be removed is uniquely determined by the rule. It can easily be proved that, if this criterion is applied, cycling can never occur, so that the procedure will always lead to an optimal solution in a finite number of steps[2,3]. Cycling is not likely to occur even if some other criterion is used, e.g., if the variable to be replaced is picked at random or designated as the one that has the lowest subscript; but the present rule of thumb, which provides an absolute guarantee, is nearly as easy to apply since a decision is usually reached in the P_1 or the P_2 column.

F. Procedure for Solving Transportation Problems

1. Like all other linear programming problems, the transportation problem can be solved by the simplex method, and the calculations are particularly simple because of the highly special nature of the problem. The coefficients in the P_j columns of the initial simplex tableau are ones and zeroes because the restrictions are mere summations of the variables; in the subsequent tableaux, no other values than 0, +1, and −1 can

---

1 *Exercise*: Explain why.

2 The rule, which is due to A. Charnes, is derived from a modified simplex procedure which makes is possible to steer clear of degenerate solutions by slightly *perturbing* the problem. In the initial tableau, each element $\bar{x}_i$ of the P_0 column is replaced by the polynomial

$$\bar{x}_i + x_{i1}\varepsilon + x_{i2}\varepsilon^2 + \ldots + x_{in}\varepsilon^n$$

where ε is a small positive, unspecified number. The coefficients of the powers of ε are those occurring in the same row in the P_j columns, and the same will be true in each of the subsequent tableaux if the expanded problem is solved by the usual simplex procedure. At no stage of the computations can degeneracy occur; the limits will be polynomials of the form

$$\frac{\bar{x}_i}{x_{ik}} + \left(\frac{x_{i1}}{x_{ik}}\right)\varepsilon + \left(\frac{x_{i2}}{x_{ik}}\right)\varepsilon^2 + \ldots$$

and no two such polynomials can be identical as the rows of the tableau will always be different. Since the polynomials are dominated by the lower powers of ε, we can stop calculating the coefficients as soon as the tie is broken; in the example above only the coefficients of the first power, the x_{i1}/x_{ik}, were calculated (=7/2 and 1/2). When degeneracy has thus been evaded, the procedure will lead to an optimal solution in a finite number of steps, and the optimal solution to the original problem is obtained by setting $\varepsilon=0$. In practice the terms involving powers of ε need never be entered in the P_0 column, as the coefficients appear in the P_j columns.

3 *Exercise*: Solve the numerical problem above, as expanded with powers of ε. (Write the complete polynomials in the P_0 column throughout the calculations.)

appear in the columns, as can easily be shown by an example. Furthermore, the particular structure of the problem makes it possible to find an initial basis of structural variables so that artificial variables can be dispensed with, and thus to save a number of steps.

The procedure may be illustrated by a numerical example. Assume that we have the following transportation problem[1]:

x_{ij}	Destination no. ($j=$)					Total
	1	2	3	4	5	
Origin 1	x_{11}	x_{12}	x_{13}	x_{14}	x_{15}	6
no. ($i=$) 2	x_{21}	x_{22}	x_{23}	x_{24}	x_{25}	6
3	x_{31}	x_{32}	x_{33}	x_{34}	x_{35}	3
Total	4	2	2	4	3	15

The unknowns, the x_{ij}, are here measured in tons per day. The corresponding cost coefficients c_{ij}, in $ per ton shipped from the ith origin to the jth destination, are as follows:

c_{ij}	$j=$				
	1	2	3	4	5
$i=$ 1	3	1	2	2	2
2	2	3	4	1	2
3	1	2	3	2	3

The simplex tableau will have $3 \cdot 5 = 15$ structural columns, and it takes $3+5-1=7$ variables to form a basis since, as we have seen, any one of the eight equations can be derived from the others.

A basis composed of 7 structural variables is easily found by the so-called "*Northwest Corner Rule*". The procedure starts from a table like the first one above, only with blank spaces for the x_{ij}. The variable corresponding to the northwest corner of the table, x_{11}, is set $=4$; this is the largest value compatible with the two equations in which x_{11} occurs, those referring to the first row and the first column. Thus the requirements at the first destination are satisfied by shipments from the first origin and all other entries in the first column will be zero. $6-4=2$ units are still available at the first origin, to be delivered at the second destination so that we have $x_{12}=2$. The first row equation and the first two column equations are now satisfied, so we move on to x_{23}, which is set equal to 2, the smaller of the numbers 2 and 6. Proceeding in this fashion, we end up with the number 3 in the southeast corner, as shown in the following table.

---

1 The problem is borrowed from A. Bordin, op. cit., p. 47.

x_{ij}		Destination no. ($j=$)					Total
		1	2	3	4	5	
Origin no. ($i=$)	1	4	2	0			6
	2			2	4		6
	3				0	3	3
Total		4	2	2	4	3	15

The procedure will always work out, because total shipments from origins are equal to total receipts at destinations. The path from the northwest to the southeast corner involves 4 horizontal and 2 vertical moves, i.e., it covers 7 cells of the table — in the general case, $m+n-1$ cells — and thus provides a basic feasible solution, although some of the x_{ij} along the path may be zero. In the present case, two of the seven basic variables are zero so that the solution provided by the rule is degenerate[1].

The coefficients of the first simplex tableau are determined by solving the equations for this initial set of basic variables, which is an easy matter because of the special nature of the equations. Furthermore, the procedure may be shortened by numbering origins and destinations in such a way as to obtain a suitable starting basis by the northwest corner rule.

2. However, the transportation problem can be solved more easily by a special procedure which is in effect a simplified version of the simplex method[2]. The procedure is based upon the fact that, once an initial basic solution has been determined by applying the northwest corner rule to an x_{ij} table, the second basic solution can be derived by a simple adjustment of table, and so forth.

In our numerical example, the first basic solution is represented by the seven numbers entered in the x_{ij} table above. The blank spaces correspond to the non-basic variables; one of them is to replace a basic variable at the next step. Let us explore the effects of increasing the value of, say, x_{21} by an unspecified positive amount x, while still keeping the other non-basic variables at zero. The adjustments made necessary by the increase in x_{21} are displayed in the following table.

	1	2	3	4	5	
1	$4-x$	2	$0+x$			6
2	$+x$		$2-x$	4		6
3				0	3	3
	4	2	2	4	3	15

1 Instead of writing $x_{13}=0$, we might have gone round by x_{22}; similarly, x_{25} might have been chosen as a basic variable instead of x_{34}. Indeed, any two blank entries in the table (i.e., zero variables) may formally be considered as representing basic variables together with the five variables that are positive in the basic solution. Cf. Ch. V, E, above.

2 See A. Charnes and W. W. Cooper (1954).

Writing $+x$ in the cell of x_{21}, some other element in the same row must be reduced by the same amount since otherwise the corresponding equation would be violated. Thus the value of x_{23} becomes $2-x^1$. In order for the equation of the third column to be satisfied, the value of x_{13} will have to be increased from 0 to x^2. After finally adjusting x_{11}, all of the equations will again be satisfied. The adjusted solution will be feasible so long as none of the variables are negative; the upper limit to x is the point where another variable first becomes zero, in the present case $x=2$. This point represents a new basic solution where x_{21} has replaced x_{23} as a basic variable.

Whether the introduction of x_{21} will improve the solution, is quite another question, which will depend on the sign of the simplex coefficient $z_{21}-c_{21}$. The table above shows how the variables are affected by the introduction of a positive x_{21} ($=x$), so the corresponding effect on the cost function is easily calculated. The net increase in total cost will evidently be

$$c_{21}x - c_{23}x + c_{13}x - c_{11}x,$$

and the simplex coefficient, i.e., the *net saving in cost per unit of* x_{21}, will therefore be

$$z_{21}-c_{21}=(c_{23}-c_{13}+c_{11})-c_{21}=3.$$

In a similar fashion, we can trace the effects on the solution of introducing any other non-basic variable into the basis, and compute its simplex coefficient. The results are displayed in the following table, which may be thought of as a condensed version of the simplex tableau for the first step.

	i	j 1	2	3	4	5	Total
I	1	**4**	2	0	−3	−2	6
	2	3	0	**2**	4	0	6
	3	5	2	2	0	**3**	3
	Total	4	2	2	4	3	15

The bold-faced entries are the values of the basic variables in the solution; all other cells are filled in with the simplex coefficients of the corresponding variables. The largest (most positive) simplex coefficient, indicated by italics, is $z_{31}-c_{31}=5$, hence x_{31} is the variable to be *introduced* into the basis. The variable to be *replaced* is the one which first becomes zero

---

1 *Exercise:* Why could we not have adjusted x_{22} or x_{24} instead of x_{23}?

2 We cannot choose to adjust x_{33} instead of x_{13}; we are examining the effects on *basic* variables only.

for increasing x_{31}; in the present case, since for $x_{31}=x$ we have $x_{34}=0-x^1$, the effective upper limit to x_{31} will be zero, so that x_{31} merely replaces x_{34} as a zero variable in the basis. In other words, one degenerate solution is replaced by another; in the new tableau we write a (bold-faced) zero in the cell of x_{31}, whereas the cell of x_{34} is now left blank. The values of all other basic variables are transferred unchanged since the correction term x is zero, and the new simplex coefficients are then calculated.

Proceeding in this fashion, we get the following tableaux:

	i	j 1	2	3	4	5	Total
II	1	4	2	0	−3	3	6
	2	3	0	2	4	5	6
	3	**0**	−3	−3	−5	3	3
III	1	2	2	2	2	3	6
	2	−2	−5	−5	4	2	6
	3	2	−3	−3	0	1	3
IV	1	1	2	2	−1	1	6
	2	*1*	−2	−2	4	2	6
	3	3	−3	−3	−3	−3	3
V	1	−1	2	2	−1	2	6
	2	1	−2	−2	4	1	6
	3	3	−2	−2	−2	−2	3
Total		4	2	2	4	3	15

In Tableau V, no positive simplex coefficients appear, so the corresponding basic solution—which is not degenerate—represents an optimal pattern of shipments[2]. Multiplying the bold-faced values by the corresponding c_{ij} and adding, the total cost of transportation is found to be $g=21$[3].

The optimal combination of routings and the values of the positive x_{ij} are shown in Fig. 10.

When a transportation problem is solved by this method, it is practical to put the values of the basic variables in circles. The correction

---

[1] *Exercise:* Show this by adjusting the values of the basic variables in the fashion demonstrated above, with x_{31} as the new basic variable ($=x$).

[2] If some of the simplex coefficients in the tableau had been zero, the solution would still have been optimal, but not unique.

[3] *Exercise:* Solve the problem in five steps by the general simplex method, omitting the equation referring to the third origin, and using the same initial basis as above. Show that the calculations correspond to those above, step for step and coefficient for coefficient.

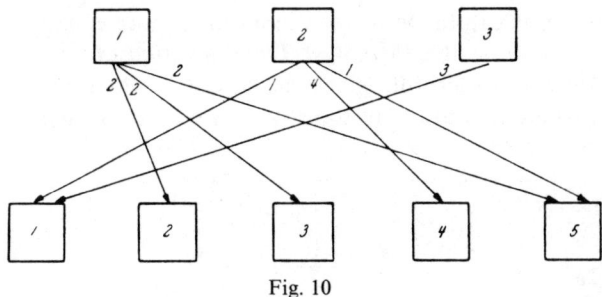

Fig. 10

terms $+x$ and $-x$ need not be written explicitly in the circled cells; the basic variables to be affected by an increase in x_{ij} are easily traced by mere inspection, moving horizontally from the cell of x_{ij} to a circle in the same row, then moving vertically to another circle, and so forth in alternating directions following the shortest possible path — until we are back in the cell of x_{ij}[1]. The circles at which the direction is changed represent the basic variables whose values are affected and whose c_{ij} will thus contribute with alternating signs to the simplex coefficient[2]. (The calculations can be facilitated by entering the c_{ij} coefficients at the top left-hand corners of the corresponding cells in each tableau.) The variable to be *replaced* will be designated by the smallest of those circled values along the path corresponding to the new basic variable that give *positive* contributions to the simplex coefficient[3]. This value represents the value of the *new* basic variable in the next tableau, i.e., the value of x. The new basic solution can easily be found by inspection, taking the rim requirements into account.

The fact that the first two basic solutions in the example were degenerate did not cause any difficulties; the required number of "stepping stones" was provided by circling a suitable number of zeroes, and there never was any doubt as to which variable was to be replaced in the basis. When difficulties of the latter nature arise, they can easily be overcome by a slight modification of the procedure resembling the one used in evading degeneracy in the general linear programming problem[4].

3. A weak point of the transportation method as described above is the rather cumbersome procedure for calculating the simplex coefficients.

---

[1] Following the shortest possible path does not imply that the direction is changed as soon as a circle is reached. For example, in calculating $z_{21} - c_{21}$ in the first tableau above, the circled value of x_{12} had to be skipped over. (*Exercise:* Explain why.)

[2] These circles may be thought of as stepping stones used in exploring the vicinity of x_{ij}, hence the method is known as the *"stepping stone method"*.

[3] *Exercise:* Explain why.

[4] Cf. A. Charnes and W. W. Cooper (1954), pp. 60ff.

These calculations can be performed in a much more efficient way, using a procedure known as the "*Modified Transportation Method*"[1].

The northwest corner rule gives the initial basic solution shown in *Tableau I a*, where a_i and b_i are the row and column sums:

	x_{ij}	1	2	3	4	5	a_i
	1	4	2	0			6
Ia	2		–	2	4		6
	3				0	3	3
	b_j	4	2	2	4	3	15

The solution is the same as that given (in bold-face type) in Tableau I above. The next step is to calculate the simplex coefficients; we proceed as follows:

(i) In a cost table (c_{ij}), fill in the cells corresponding to the basic solution[2]; add an additional column c_i and an additional row d_j:

c_{ij}	1	2	3	4	5	c_i
1	3	1	2			c_1
2			4	1		c_2
3				2	3	c_3
d_j	d_1	d_2	d_3	d_4	d_5	

(ii) Determine the c_i and d_j such that $c_i + d_j = c_{ij}$ for all c_{ij} in the table. This can be done in many ways. For example, start by setting $c_1 = 0$; then we must have $d_1 = 3$ if $c_1 + d_1$ is to be $= c_{11} = 3$, and similarly $d_2 = 1$, $d_3 = 2$. It follows that $c_2 = 2$, and so forth. The results are entered in the c_i column and the d_j row.

(iii) Fill in the empty cells of the table with the values $c_i + d_j - c_{ij}$, the difference between the "fictitious cost" $c_i + d_j$ and the true cost per ton for this route, c_{ij}. As will be shown below, the fictitious cost is equal to z_{ij}; the difference $z_{ij} - c_{ij}$ is the net saving in cost per unit of x_{ij} if route (i,j) is introduced, i.e., $c_i + d_j - c_{ij}$ is the simplex coefficient of x_{ij}. In this way we get *Tableau I b*:

---

[1] See S. Vajda (1962), pp. 3ff.

[2] The entries are shown in bold-faced type. For hand solution they may be put in circles.

	c_{ij}	1	2	3	4	5	c_i
	1	**3**	**1**	**2**	0−1−2	0+0−2	0
I b	2	2+3−2	2+1−3	**4**	**1**	2+0−2	2
	3	3+3−1	3+1−2	3+2−3	**2**	**3**	3
	d_j	3	1	2	−1	0	

Tableau Ia—b corresponds to, and conveys the same information as, Tableau I above. We can now proceed to the next basic solution, using the familiar rules for selecting a new basic variable. The modified method differs from the original transportation method only in the way the simplex coefficients are calculated at each step.

It is easy to see why $c_i + d_j$ must be equal to z_{ij}. The simplex coefficient $z_{21} - c_{21}$, for example, was calculated in Tableau I as

$$z_{21} - c_{21} = (c_{23} - c_{13} + c_{11}) - c_{21}$$

where the cost coefficients entering into z_{21} correspond to basic variables — viz., those basic variables which are affected if x_{21} becomes positive, as indicated by $+x$ or $-x$ in the adjustment table — and the fictitious costs were determined precisely such as to satisfy $c_{ij} = c_i + d_j$. We then have

$$z_{21} = c_{23} - c_{13} + c_{11} = c_2 + d_3 - c_1 - d_3 + c_1 + d_1$$
$$= c_2 + d_1.$$

This result holds for every z_{ij}. Any two successive terms in the expression for z_{ij} have either a row or a column in common and have opposite signs so that the corresponding c or d cancels out; all that remains is $c_i + d_j$, the fictitious cost.

Proceeding from Tableau Ia—b, the largest positive simplex coefficient — in this case $z_{31} - c_{31} = c_3 + d_1 - c_{31} = 5$ — points to x_{31} as the variable to come into the basis. The variable to be replaced is determined by the usual adjustment procedure, which can now be carried out within Tableau Ia. The new basic solution, where x_{31} has replaced x_{34}, is set up in a *Tableau IIa* (not shown here but analogous to Tableau Ia).

The simplex coefficients corresponding to the new basis are calculated in the same way as used for Tableau Ib; as the basis has been changed, we cannot use the same values for the c_i and the d_j and *new* values have to be calculated, again setting e.g. $c_1 = 0$. The results are entered in a *Tableau IIb*, arranged like Tableau Ib.

Proceeding in this fashion until we get to a basis with no positive simplex coefficients, the optimal solution emerges in Tableau Va—b, corresponding to Tableau V above.

The calculations can be arranged as follows:

	x_{ij}	1	2	3	4	5	a_i	c_{ij}	1	2	3	4	5	c_i
	1	4−x	2	0+x			6	1	**3**	**1**	**2**	−3	−2	0
	2			2−x	4+x		6	2	3	0	**4**	**1**	0	2
I	3	+x			0−x	3	3	3	**5**	**2**	**2**	**2**	**3**	3
	b_j	4	2	2	4	3	15	d_j	3	1	2	−1	0	
	1	4−x	2	0+x			6	1	**3**	**1**	**2**	−3	3	0
	2			2−x	4	+x	6	2	3	0	**4**	**1**	5	2
II	3	0+x			3−x	3	3	3	**1**	−3	−3	−5	**3**	−2
	b_j	4	2	2	4	3	15	d_j	3	1	2	−1	5	
	1	2−x	2	2		+x	6	1	**3**	**1**	**2**	2	*3*	0
	2				4	2	6	2	−2	−5	−5	**1**	**2**	−3
III	3	2∣x				1−x	3	3	**1**	−3	−3	0	**3**	−2
	b_j	4	2	2	4	3	15	d_j	3	1	2	4	5	
	1	1−x	2	2		1+x	6	1	**3**	**1**	**2**	−1	**2**	0
	2	+x			4	2−x	6	2	*1*	−2	−2	**1**	**2**	0
IV	3	3					3	3	**1**	−3	−3	−3	−3	−2
	b_j	4	2	2	4	3	15	d_j	3	1	2	1	2	
	1		2	2		2	6	1	−1	**1**	**2**	−1	**2**	0
	2	1			4	1	6	2	**2**	−2	−2	**1**	**2**	0
V	3	3					3	3	**1**	−2	−2	−2	−2	−1
	b_j	4	2	2	4	3	15	d_j	2	1	2	1	2	

To avoid unnecessary repetitions a simplified tabular form like the following may be applied.

		a					b				
		1	2	3	4	5	1	2	3	4	5
	1	4−x	2	0+x			**3**	**1**	**2**	−3	−2	0
	2			2−x	4+x		3	0	**4**	**1**	0	2
I	3	+x			0−x	3	**5**	**2**	**2**	**2**	**3**	3
							3	1	2	−1	0	
	1	…					…				…	
II	2											
	3											
⋮							…					

4. The amount of computational work involved in solving a transportation problem can be further reduced by searching for an initial basic solution which is as close as possible to the optimal pattern. The result obtained by applying literally the northwest corner rule depends on the order in which the rows and the columns are arranged in the transportation table, and this ordering is completely arbitrary. With the ordering used above, the problem was solved in five steps, but if we had happened to write the rows in the order (3, 2, 1) and the columns in the order (1, 4, 5, 2, 3), the northwest corner rule would immediately have produced the initial basic solution of Tableau V so that the problem would have been solved in a single step[1]. More efficient methods of picking a good initial solution are clearly called for.

An obvious procedure would be to *occupy the cheapest routes first*. In the example above, there are three routes having the cost coefficient $c_{ij}=1$; each of the corresponding cells in the transportation table should then be filled in with the largest number compatible with the restrictions. These entries are indicated by the top index (^{1}) in the following table. Having thus disposed of the third row and the second and fourth column, which can now be deleted, the procedure is repeated in the reduced table, occupying as many "second-best" cells (those with $c_{ij}=2$) as we can within the restrictions. However, the result will depend on the order in which these cells are occupied. If we fill in cells (2,1), (1,3), and (1,5) in that order, we get the entries marked by (^{2}) in the table; all that remains is to set $x_{25}=1$:

	2^1	2^2		2^2	6
1^2			4^1	1^3	6
3^1					3
4	2	2	4	3	15

The initial solution thus obtained is seen to be identical with the solution in Tableau Va above. Calculating the simplex coefficients, using fictitious costs, we get Tableau Vb and the problem has been solved in a single step.

Had we filled in the second-best routes in the reverse order, we would have got an initial solution which is several steps removed from the optimal one[2]:

	2^1	1^2		3^2	6
1^2		1^3	4^1		6
3^1					3
4	2	2	4	3	15

---

1 *Exercise:* Show this.
2 *Exercise:* Solve the problem, starting from this initial solution.

A related procedure, which leaves no doubt as to the order in which the cells are to be filled in, is as follows.

The cheapest route in the *first column* is $(3, 1)$; set $x_{31} = 3$, the largest possible shipment. This disposes of the third row. The second-best route in the first column is $(2, 1)$; therefore, set $x_{21} = 1$ so that we can delete the first column. Within the reduced problem, repeat the procedure in the *second column*, and so forth. The resulting solution, shown in the following table, is seen to be that of Tableau Va so that the problem has been solved in a single step[1,2].

	2^3	2^4		2^6	6
1^2			4^5	1^7	6
3^1					3
4	2	2	4	3	15

5. The transportation method has a wider field of application than might be supposed at first sight. By applying various devices, certain classes of apparently quite different linear programming problems can be transformed into models which are formally identical with the transportation problem, and which can therefore be solved by the much simpler transportation method.

For example, the inventory problem of Ch. IV, C can be formally regarded as the planning of deliveries from eight origins (two shifts in each of the four production periods) to five destinations, viz. the four sales periods and a dummy destination absorbing unutilized capacities such as to balance the problem[3]. With an initial solution corresponding to that given in Ch. IV,C inserted, the transportation table looks as follows.

Produced in quarter	shift	x_{ij}	1	2	3	4	5 (dummy)	a_i
1	1	1	550.5					550.5
1	2	2	161.5				291.5	453.0
2	1	3		550.5				550.5
2	2	4		27.5			425.5	453.0
3	1	5			547.0	3.5		550.5
3	2	6				209.0	244.0	453.0
4	1	7				550.5		550.5
4	2	8				453.0		453.0
		b_j	712.0	578.0	547.0	1,216.0	961.0	4,014.0

1 *Exercise:* Solve the problem by applying a similar procedure to the *rows* instead of the columns.

2 *Exercise:* Solve the transportation problems in Ch. IV, D, by the modified transportation method.

3 Cf. S. Vajda (1962), Ch. IV.

The costs of production are 60.62 D. cr. per unit produced in the first shift and 73.66 per unit produced in shift 2, while storage costs ("transportation" costs in time) are 1.01 per unit per quarter. Thus, for example, a unit produced in shift 1 in the second quarter ($i=3$) and sold in the fourth quarter ($j=4$) will cost $c_{34}=60.62+2\times1.01=62.64$. Deliveries backwards in time are given a prohibitive cost coefficient M such as to ensure that the corresponding x_{ij} will be zero in the solution; the cost coefficients in the dummy column are set $=0$ since no costs are associated with producing and selling nothing. The cost table then looks as follows:

c_{ij}	1	2	3	4	5	c_i
1	60.62	61.63	62.64	63.65	0	0
2	73.66	74.67	75.68	76.69	0	13.04
3	M	60.62	61.63	62.64	0	0
4	M	73.66	74.67	75.68	0	13.04
5	M	M	60.62	61.63	0	0
6	M	M	73.66	74.67	0	13.04
7	M	M	M	60.62	0	−1.01
8	M	M	M	73.66	0	12.03
d_j	60.62	60.62	60.62	61.63	−13.04	

Calculating the fictitious costs corresponding to the solution, we find that $z_{ij}-c_{ij}=c_i+d_j-c_{ij}<0$ for the non-basic routes (except that $z_{63}-c_{63}=0$) so that the solution is optimal (but not unique)[1].

---

[1] *Exercise:* Show that this solution emerges as the initial basic solution if the column procedure described above is applied (fill in the first column of the transportation table, starting with the cheapest route, etc.). The zeroes in the dummy column, of course, do not count as cheapest routes; the procedure is applied to the structural columns only and the entries in the dummy column are then determined residually.

VI. Duality in Linear Programming

A. The Duality Theorem

Linear programming models possess the interesting property of forming pairs of symmetrical problems. To any maximization problem corresponds a minimization problem involving the same data, and there is a close correspondence between their optimal solutions. The two problems are said to be "*duals*" of each other.

As an example let us take the problem (12) of Ch. II:

$$\begin{align} 2x_1 \quad\quad\quad +4x_3 &\geq 5 \\ 2x_1+3x_2+1x_3 &\geq 4 \\ 4x_1+2x_2+3x_3 &= g = \text{minimum}. \end{align} \tag{1}$$

The dual of (1) is the following problem in two structural variables, y_1 and y_2:

$$\begin{align} 2y_1+2y_2 &\leq 4 \\ 3y_2 &\leq 2 \\ 4y_1+1y_2 &\leq 3 \\ 5y_1+4y_2 &= f = \text{maximum}. \end{align} \tag{2}$$

The coefficients of the two problems are the same, namely

2	0	4	5
2	3	1	4
4	2	3	

except that the rows and the columns have been interchanged, i.e., the coefficients of (1) are read off the table row-wise, those of (2) columnwise. The right-hand terms of one problem become the coefficients in the preference function of the other; furthermore, whereas (1) is concerned with minimizing a linear function subject to inequalities of the type $\geq$, (2) is a maximization problem involving inequalities of the reverse type.

The optimal basis in (1) is (x_2, x_3); solving for these variables in terms of x_1 and the slack variables we get

$$x_2 = \frac{11}{12} - \frac{1}{2}x_1 - \frac{1}{12}x'_1 + \frac{1}{3}x'_2$$

$$x_3 = \frac{5}{4} - \frac{1}{2} x_1 + \frac{1}{4} x'_1 \qquad (3)$$

$$g = \frac{67}{12} + \frac{3}{2} x_1 + \frac{7}{12} x'_1 + \frac{2}{3} x'_2$$

where the simplex coefficients are all positive so that the optimal basic solution is obtained by setting $x_1 = x'_1 = x'_2 = 0$. In (2), the optimal basis turns out to be (y_1, y_2, y'_1) where y'_1 is the slack variable of the first equation, since the solution

$$y_1 = \frac{7}{12} + \frac{1}{12} y'_2 - \frac{1}{4} y'_3$$

$$y_2 = \frac{2}{3} - \frac{1}{3} y'_2$$

$$y'_1 = \frac{3}{2} + \frac{1}{2} y'_2 + \frac{1}{2} y'_3 \qquad (4)$$

$$f = \frac{67}{12} - \frac{11}{12} y'_2 - \frac{5}{4} y'_3$$

satisfies the simplex criterion.

Comparing (3) with (4), there appears to be a complete correspondence between the two solutions; all coefficients of (3) occur in (4) and vice versa[1]. In particular, we observe that

(i) *the maximum value of f is equal to the minimum value of g,* and that

(ii) *in the optimal basic solutions, the value of any variable in (1) is numerically equal to the simplex coefficient of the corresponding variable in (2) and vice versa.* (By "corresponding variables" we mean the pair (x_i, y'_i) or (y_j, x'_j)). As a corollary, it follows from (ii) that, if the optimal basis in one of the two problems is known — e.g., (x_2, x_3) — then the corresponding variables in the other problem, in this case y'_2 and y'_3, will be zero in the optimal solution so that the optimal basis will consist of the remaining variables $(y_1, y_2,$ and $y'_1)$.

This result, known as the *Duality Theorem,* can be shown to hold in the general case[2,3], except that it is not possible to establish perfect

---

1 *Exercise:* Demonstrate the same correspondence between the final simplex tableaux for (3) and (4). The former is displayed on p. 83.

2 A general proof of the theorem is given in the Appendix, D.

3 *Exercise:* Formulate the dual of the trivial problem

$$a_1 x_1 + a_2 x_2 \leq b$$
$$c_1 x_1 + c_2 x_2 = f = \text{maximum}$$

(assuming that x_1 constitutes the optimal basis), and demonstrate the validity of the duality theorem in this special case. Generalize to the case of two inequalities in the same two structural variables.

symmetry when one of the problems involves equations in the structural variables[1,2].

Thus, if one of the two problems is solved, the optimal solution to the other follows as a by-product; for example, in the final simplex tableau of the "primal" problem (1), the solution to its dual (2) appears in the $z_j - c_j$ row, and the P_0 column gives the simplex coefficients of the non-basic variables in (2). This means that we have a choice of two alternative ways of solving a concrete linear programming problem, a fact which can sometimes be exploited to reduce the computational labour since, as a rule, it will be easier to solve that problem which involves fewer restrictions because the expected number of iterations will be smaller.

B. Economic Interpretation of the Dual

1. The phenomenon known as duality is not only a formal property of the mathematical programming model. As we shall see, the dual problem will always have a meaningful interpretation according to the economic nature of the original problem.

Suppose, for example, that (1) is a blending problem of the diet type, concerned with finding the cheapest possible mixture of ingredients (inputs) which will contain the prescribed minimum amounts of the two constituents (the "outputs" of the blending process). The x's are the *quantities* of the ingredients, with prices 4, 2, and 3 respectively.

Now, what is the interpretation of the dual problem (2)? We first observe that the dual variables y_1 and y_2 have the dimension of *prices*; the preference function $f = 5y_1 + 4y_2$ is measured in \$ (its maximum value being equal to the minimum value of total cost g) and the coefficients 5 and 4 are measured in physical units of output, hence the dimension of y_1 and y_2 must be \$ per unit of output. Consequently the values of y_1 and y_2 may be thought of as a set of prices associated with the two outputs — not market prices, of course, but *imputed* values, so-called "*shadow prices*". What is the economic interpretation of the shadow prices?

By the Duality Theorem, the optimal values of the y's, $\bar{y}_1$ and $\bar{y}_2$, correspond to the simplex coefficients in the original problem, cf. (3) and (4). For example, $\bar{y}_1$ will be $= 7/12$, corresponding to the simplex coefficient of x'_1. Now, by (3), this coefficient is to be interpreted as the net increase in total cost of ingredients caused by increasing x'_1 from 0 to 1, i.e., the cost of producing an additional unit of the first output.

---

[1] *Exercise:* Explain why. (Hint: Compare the numbers of variables and simplex coefficients in the two problems.)

[2] Cf. R. L. Ackoff (1961), pp. 118f.

A similar reasoning applies to $\bar{y}_2$. In other words, the shadow prices provide a valuation of the outputs based upon marginal cost considerations.

Using these internal prices, we may calculate the total imputed value of the quantities of outputs produced by a unit of each particular input, and compare with the actual cost, i.e., with the market price of the ingredient, to obtain a criterion for deciding which ingredients can be uset at a profit.

This is precisely what is implied in the solution of the dual problem (2). It follows directly from the Duality Theorem that, when the least-cost combination — the optimal solution to (1) — consists of x_2 and x_3, whereas x_1 is $=0$, then the slack variables y'_2 and y'_3 in the dual problem (2) will be zero, whereas y'_1 will be positive. In other words, those equations of (2) which correspond to the optimal ingredients of (1) will express an equality between total imputed value contributed per unit of input, and the actual price of this unit:

$$3\,\bar{y}_2 = \qquad 3 \cdot \frac{2}{3} = 2$$

$$4\,\bar{y}_1 + 1\,\bar{y}_2 = 4 \cdot \frac{7}{12} + 1 \cdot \frac{2}{3} = 3,$$

whereas the contribution to total value of output made by a unit of x_1 is less than its price, as demonstrated by the corresponding inequality:

$$2\,\bar{y}_1 + 2\,\bar{y}_2 = 2 \cdot \frac{7}{12} + 2 \cdot \frac{2}{3} = \frac{5}{2} < 4,$$

so that the ingredient will not be used in the optimal combination.

Thus the shadow prices, as determined by solving the dual problem, provide a criterion of optimality which is equivalent to determining the optimal combination of x's directly from (1). In other words, the problem of optimal allocation can be formulated and solved alternatively in terms of prices or quantities[1]. This is the economic content of the duality property of the mathematical model.

---

[1] A concrete example of shadow prices being applied by a company to solve a practical problem of production planning is the ice cream problem of Ch. III, cf. p. 24 above. To compare several ingredients, all of which contain fat and serum solids, a price is assigned to one of these constituents, e.g., to serum solids, so that the "net cost of fat" in each ingredient can be calculated as its market price minus the value of the serum solids content. The choice is then based on the lowest net price per unit of fat. *Exercise:* Show that this procedure — as described on p. 24 — is in effect an application of the Duality Theorem when the value assigned to serum solids is determined by solving the dual, and solve the complete problem (p. 19) by means of the shadow prices. (Hint: The shadow prices are equal to the simplex coefficients on p. 20. The optimal solution of the original problem remains the same if the equations are transformed into inequalities, so there is complete symmetry between the two problems).

For the production programme as a whole, the Duality Theorem establishes an equality between total cost of inputs and total imputed value of outputs in the optimal combination:

$$g_{\min} = 2\,\bar{x}_2 + 3\,\bar{x}_3 = 67/12 = 5\,\bar{y}_1 + 4\,\bar{y}_2 = f_{\max}.$$

2. The interpretation of the dual of an economic *maximization* problem is, of course, quite analogous. Let us take the problem (2) as our starting point, now interpreting it as a problem concerned with optimal utilization of three machines whose capacities — measured in machine hours available per period — are limited. The variables y_1 and y_2 are the quantities to be made per period of the company's two products, and the coefficients of the preference function are unit gross profits (selling prices minus variable costs).

(1) is the corresponding dual problem, where the x's must now be interpreted as shadow prices associated with the services of the fixed factors, measured in $ per machine hour. What is now to be minimized in (1), is total imputed cost of using the machines. In a problem of this type, only such products should be made for which gross profit just covers total imputed cost of machine time used per unit of product; whenever unit profit falls short of total machine cost as valuated on the basis of the shadow prices, the corresponding variable is zero in the optimal solution to (2). (In the present example, both products can be made at a profit.) The total profit associated with the optimal production programme will be equal to the total imputed value of machine hours used in producing the programme.

For example, the shadow price $\bar{x}_3 = 5/4$ corresponds to the simplex coefficient of y'_3 in the profit function f, i.e., the services of the third fixed factor are valued on the basis of their marginal contribution to total profit[1]. The shadow prices bear no direct relation to the market prices at which the machines were originally purchased; these "historical" costs (which give rise to fixed charges in each period) are left out of the picture since they are independent of the production programme chosen and thus cannot affect the optimal solution.

Many business firms actually base their choice of a production programme on cost estimates which include internal valuations of the costs associated

---

[1] It follows that the shadow price represents the highest price which the company would be willing to pay for an additional machine hour, or in general, for an additional unit of a scarce input.

If a fixed factor is not fully utilized in the optimum (such as the first machine in the present example, since $\bar{y}'_1 > 0$), it follows from the Duality Theorem that the corresponding shadow price will be zero ($\bar{x}_1 = 0$), a result which is confirmed by common sense; there is no reason why the company should be willing to pay for additional machine hours when the machine is idle part of the time.

with using the fixed factors. Such a procedure will not lead to maximum profit unless the valuations agree with the shadow prices of the linear programming model[1].

C. The Dual Simplex Method

It follows from the Duality Theorem that a linear programming problem can be solved by applying the simplex method either to the problem as it stands (the primal) or to its dual.

It is sometimes convenient to solve the problem in a third way, namely by performing the operations of the dual solution within the framework (i.e., the simplex tableaux) of the primal problem. We shall demonstrate this procedure in a simple example.

Let the *primal* problem be

$$2 x_1 \geq 4$$
$$3 x_1 \geq 2$$
$$12 x_1 = g = \text{minimum}.$$

To get a feasible initial solution, introduce slack variables x'_1 and x'_2 and artificial variables x'_3 and x'_4. The problem is then solved in three steps:

		x_0	12 x_1	0 x'_1	0 x'_2	M x'_3	M x'_4
I	x'_3	4	2	-1	0	1	0
	x'_4	2	3	0	-1	0	1
		$6M$	$5M-12$	$-M$	$-M$	0	0
II	x'_3	8/3	0	-1	2/3	1	$-2/3$
	x_1	2/3	1	0	$-1/3$	0	1/3
		$\frac{8}{3}M+8$	0	$-M$	$\frac{2}{3}M-4$	0	$-\frac{5}{3}M+4$
III	x'_2	4	0	$-3/2$	1	3/2	-1
	x_1	2	1	$-1/2$	0	1/2	0
		24	0	-6	0	$-M+6$	$-M$

The *dual* is

$$2 y_1 + 3 y_2 \leq 12$$
$$4 y_1 + 2 y_2 = f = \text{maximum}.$$

With the slack variable y'_1 as the initial basic variable we get an optimal solution in two steps[2]:

---

1 *Exercise*: Solve the duals of the problems in Ch. IV.
2 *Exercise*: Check the correspondence between the simplex tableaux III and II′.

		y_0	y_1	y_2	y'_1
I'	y'_1	12	2	3	1
		0	−4	−2	0
II'	y_1	6	1	3/2	1/2
		24	0	4	2

The latter procedure was the more efficient.

The *dual simplex method*[1], applied to the primal, starts from a basic solution in which the *simplex coefficients* have the sign required for the optimum solution (i.e., the simplex criterion is satisfied), whereas one or more of the *basic variables* is negative. (Such a solution is said to be *dual feasible*; it corresponds to a feasible but non-optimal basic solution to the dual.) The negative variables are then eliminated one by one from the basis in a sequence of simplex tableaux. The procedure differs from the usual (primal) simplex method in that the variable to be introduced into the basis and the variable to be replaced are selected in a different way, viz. such that the successive bases correspond to those we would get by solving the dual problem.

In the primal problem, the slack variables x'_1 and x'_2 provide a negative basic solution

$$x'_1 = -4 + 2x_1$$
$$x'_2 = -2 + 3x_1$$

and the simplex coefficient (that of x_1) is seen to have the proper sign,

$$g = 0 + 12 x_1.$$

The initial simplex tableau corresponding to this solution is Tableau I'' below.

		x_0	x_1	x'_1	x'_2
I''	x'_1	−4	−2	1	0
	x'_2	−2	−3	0	1
		0	−12	0	0
II''	x_1	2	1	−1/2	0
	x'_2	4	0	−3/2	1
		24	0	−6	0

The rules by which we get to the next simplex tableau (II'') using the dual simplex method are as follows:

---

1 Cf. S. Vajda (1961), Ch. 5 and (1962), Ch. XIX.

(i) *The variable to be replaced* is now selected first; we choose the *most negative* basic variable, x'_1. (This corresponds to the choice of variable to be introduced in the dual, as designated by the most negative simplex coefficient — that of y_1 in Tableau I'; note that x'_1 and y_1 are corresponding variables.)

(ii) *The variable to be introduced* is designated by the numerically lowest of the ratios of the simplex coefficients to the corresponding numbers in the row of the new basic variable, provided that the latter numbers are negative. (Cf. the calculation of limits in the dual Tableau I'.) In this case there is only one candidate, x_1, which therefore replaces x'_1 in the basis.

Tableau I'' is now transformed into Tableau II'' row by row, applying the usual substitution and elimination rules from the simplex procedure. It follows from (ii) above that the simplex coefficients will still have the proper sign in the new tableau; therefore we have found an optimal solution when all basic variables have become non-negative. This happens already in Tableau II'', which is therefore the final tableau[1].

The simplex tableau at each step (I'' and II'') is seen to be the dual of the corresponding tableau in the solution of the dual problem (I' and II'), so the number of steps required to solve the problem by the dual simplex method is the same as that involved in solving the dual problem (in this case, two), whereas the (primal) simplex method requires three iterations and the added inconvenience of having to use artificial variables.

However, it is especially in parametric programming that the dual simplex method comes in useful, because we then have to deal with basic solutions which always satisfy the simplex criterion but become negative in some of the variables when a parameter passes a critical value[2].

---

1 Note that g, which is to be *minimized, increases* from 0 in Tableau I'' to 24 in Tableau II''. *Exercise:* Explain why.

2 See Ch. VII below.

VII. Sensitivity Analysis and Parametric Programming

A. Sensitivity Analysis

There are three groups of parameters in a linear programming problem: the "technological" coefficients, a_{ij} (representing, for example, machine time per unit of product); the constant terms on the right-hand sides of the restrictions, b_i (e.g., capacity limits); and the coefficients in the linear preference function, c_j (for example, unit profits). In practical applications of linear programming it is important to explore the *sensitivity* of the numerical solution with respect to changes in these parameters[1]. Some of them may be subject to known variations in time — prices or cost elements change, machine times are reduced and capacities increased because of rationalization or technological change, output stipulations vary from period to period, etc. — or it may not be possible to determine them exactly but only within certain intervals. (When these variations are of a random nature, the coefficients should be thought of as probability distributions rather than numbers and the problem becomes a *stochastic programming problem*.)

One way of dealing with coefficient variations is, of course, to start all over again and solve a new problem whenever one of the coefficients has changed. In many practical cases it will be useful to perform numerical experiments with the model with precisely this in view. However, when the optimum responds *discontinuously* to changes in certain strategic parameters, it is often worth while to compute, for each of these coefficients separately, the limit or limits beyond which the basis is no longer optimal (*sensitivity analysis*). This is done by treating the strategic coefficients as unspecified parameters in the calculations; those values at which one of the simplex coefficients or one of the basic variables changes sign (i.e., becomes zero) are the critical values of the parameters[2].

---

1 *Exercise:* Draw a diagram of a linear programming problem in two structural variables, and show how the solution (the optimal point) is affected by a variation in one of the parameters (one of the a_{ij}, b_i, and c_j respectively). Note that, in a bottleneck problem, the position of the optimum will not be affected by an increase in the capacity of a machine which is not fully utilized. (The corresponding slack variable will then be positive and the shadow price will be zero.)

2 Procedures for finding the critical values of the b_i and c_j ("ranging") are built into standard computer systems for linear programming.

B. A Concrete Example of Sensitivity Analysis

1. It is particularly important to calculate the critical values of the coefficients c_j since prices will, in general, be subject to more frequent variations in the short run than the other parameters of the model. We will illustrate the procedure by examining how the solution of the ice cream blending problem above (Ch. III) depends on the ingredient prices. We wish to determine the limits or intervals within which the prices must keep if the solution which we have found is to remain optimal.

The procedure runs as follows: Having solved the equation system for the basis variables x_2, x_8, x_9, and x_{10} in terms of the other variables (Ch. III, p. 19 f.), we substitute as before the solution in the cost expression $c = c_1 x_1 + c_2 x_2 + \ldots + c_{11} x_{11}$, this time, however, without specifying particular prices. This gives c as a function of the non-basic x's and of the eleven prices:

$$c = (67.394 c_2 + 3.459 c_8 + 22.483 c_9 + 1.000 c_{10})$$
$$+ (c_1 - 1.739 c_2 + 0.068 c_8) \cdot x_1 + (c_3 - 3.500 c_2 + 0.249 c_8) \cdot x_3$$
$$+ (c_4 - 3.478 c_2 + 0.221 c_8) \cdot x_4 + (c_5 - 4.339 c_2 + 0.308 c_8) \cdot x_5$$
$$+ (c_6 - 0.174 c_2 - 0.068 c_8) \cdot x_6 + (c_7 \qquad\qquad - 0.289 c_8) \cdot x_7$$
$$+ (c_{11} - 0.761 c_2 + 0.054 c_8 + 0.127 c_9 - 0.900 c_{10}) \cdot x_{11}.$$

Now the basis will represent an optimal solution if the coefficients of the x's in the cost function are non-negative:

$c_1 - 1.739 c_2 + 0.068 c_8 \geq 0$	(coefficient of x_1)	(1)
$c_3 - 3.500 c_2 + 0.249 c_8 \geq 0$	(coefficient of x_3)	(2)
$c_4 - 3.478 c_2 + 0.221 c_8 \geq 0$	(coefficient of x_4)	(3)
$c_5 - 4.339 c_2 + 0.308 c_8 \geq 0$	(coefficient of x_5)	(4)
$c_6 - 0.174 c_2 - 0.068 c_8 \geq 0$	(coefficient of x_6)	(5)
$c_7 \qquad\qquad - 0.289 c_8 \geq 0$	(coefficient of x_7)	(6)
$c_{11} - 0.761 c_2 + 0.054 c_8$ $+ 0.127 c_9 - 0.900 c_{10} \geq 0$	(coefficient of x_{11}).	(7)

This system of inequalities gives a set of sufficient (and, since the solution is not degenerate, also necessary) conditions, in terms of prices, for the given basis to be optimal. In other words, if we have a given set of market prices, we can check the optimality of the basis by making sure that the prices satisfy all of the seven inequalities, i.e., the simplex criterion. Thus, if we formulate the simplex criterion with prices as parameters, instead of using the actual prices from the beginning, we do not have to repeat the whole simplex procedure whenever prices change.

2. More specifically, the firm may be interested in examining what happens if one of the eleven prices changes while the others remain constant (*ceteris paribus*).

Let us assume, for example, that c_1 varies while c_2—c_{11} remain as specified in Ch. III. Since c_1 only occurs in the coefficient of x_1 in the cost function, we need only consider the first inequality; inserting $c_2 = 0.174$ and $c_8 = 0.165$, we get a *lower limit* for c_1:

$$c_1 - 0.302 + 0.011 \geq 0, \text{ or } c_1 \geq 0.291 \qquad \text{[from (1)]}.$$

This is the critical level of c_1. If the price rises from its initial level (0.298), or falls slightly, nothing happens, but if c_1 falls below 0.291, the combination of ingredients 2, 8, 9, and 10 will no longer be optimal. In the latter case, how will the optimum change?

The fact that it is the coefficient of x_1 which becomes negative, indicates that total cost can be lowered by substituting ingredient no. 1 for one of the four ingredients in the combination, as we should expect since no. 1 has become cheaper. Inspection of equations (8)—(11) above (Ch. III, p. 19f.) will tell which ingredient will be replaced by no. 1 in the optimal combination, and how far the substitution can go.

In the case of a change in c_2, we shall have to consider six inequalities since c_2 occurs in six of the simplex coefficients. Proceeding in the same manner as above, we then get six *upper* limits:

$c_2 \leq 0.178$ [from (1)]
$c_2 \leq 0.177$ [from (2)]
$c_2 \leq 0.176$ [from (3)]
$c_2 \leq 0.177$ [from (4)]
$c_2 \leq 0.198$ [from (5)]
$c_2 \leq 0.951$ [from (7)].

A fall in c_2 from the initial level of 0.174 does not affect the optimal combination. If c_2 rises, it will soon hit the limit 0.176 and the coefficient of x_4 becomes negative; this is the *effective* upper limit.

Similar computations for changes in the other prices give the following results:

$c_3 \geq 0.568$	$c_7 \geq 0.048$	$c_8 \geq 0.120$	$(c_8 \geq -10.796)$
$c_4 \geq 0.569$	$c_8 \geq 0.074$	$c_8 \leq 0.221$	$(c_9 \geq - 4.598)$
$c_5 \geq 0.704$	$c_8 \geq 0.117$	$c_8 \leq 0.180$	$c_{10} \leq 1.082$
$c_6 \geq 0.041$	$c_8 \geq 0.131$		$c_{11} \geq 0.498$

(where the effective price limits — i.e., the critical values beyond which our original solution is no longer optimal — are italicized). It appears that each of the prices of the ingredients which are not in the original optimum has a lower limit, as we would expect; as the price falls, sooner or later it will become economical to use it. As for the prices of the basic ingredients, c_8, like c_2, has several critical values. However, they are not all upper limits; in fact, there are five lower and two upper bounds on c_8, i.e., two effective limits. There appears to be no effective (positive) limit to c_9 — which is very natural in view of the fact that, as we have seen, liquid sugar is the only possible ingredient that can satisfy the sugar solids requirement — and c_{10} has only one upper limit.

It should be noted that not a few of the actual prices are rather close to their critical levels; very small price changes can easily upset the balance[1].

3. It should be stressed once again that the limits which we have just computed, and the results which are based on them, are valid only under strict *ceteris paribus* assumptions, i.e., for isolated changes in one of the input prices, the others remaining constant at the given original levels. If two or more prices change simultaneously, the calculations cannot be relied upon to tell what happens. For, if one price c_j changes, though this change may not itself make the original solution inoptimal, it nevertheless provides a new background against which a change in a second price c_k must be viewed.

It follows that, even if we have a set of ingredient prices which are all within the critical limits, we cannot always be sure that the combination of ingredients 2, 8, 9, and 10 is optimal, if more than one of the prices deviate from the original level. In such a case the proper procedure is to insert the actual prices for the price parameters in the cost function (i.e., in (1)—(7)) and see whether or not the simplex criterion is satisfied.

C. Parametric Linear Programming

1. By *parametric programming* is meant a more complete analysis of what happens to the solution if one of the coefficients is allowed to vary parametrically through its range of definition (e.g. from zero through all positive values). The solution will then be represented by a sequence of simplex tableaux, each one being optimal for a certain interval of the parameter in question.

In principle it is possible to explore the effect of simultaneous variation of several parameters, but the more coefficients we treat as parameters, the more complex the analysis becomes. In the following we shall stick to single-parametric cases.

Let us consider problem (12) in Ch. II (p. 10),

$$2x_1 \qquad +4x_3 \geq 5$$
$$2x_1+3x_2+ \ x_3 \geq 4$$
$$4x_1+2x_2+3x_3 = g = \text{minimum},$$

---

[1] A slight fall in c_6 will give a new combination in which x_{14} — the quantity of water to be added — is negative. In such cases the problem must be reformulated, using five effective requirements (viz., eqs. (1)–(4) and (7) of Ch. III) instead of four.

which may be interpreted as a diet problem. The optimal solution was found to be $x_2 = 11/12$, $x_3 = 5/4$, $g = 67/12$.

2. Now let the coefficient a_{11} ($=2$, the coefficient of x_1 in the first restriction) be subjected to variation. This can be expressed either by setting $a_{11} = 2 + t$, where the parameter t is an additive change in the coefficient, or by a multiplicative change ($a_{11} = 2 \cdot t$); or the parameter t may represent the coefficient itself ($a_{11} = t$). It does not matter which procedure we choose; the results will be the same. We choose to set $a_{11} = 2 + t$ in the first restriction. The simplex tableau corresponding to the basis (x_2, x_3) then becomes

			0	4	2	3	0	0
			x_0	x_1	x_2	x_3	x'_1	x'_2
I	2	x_2	$\dfrac{11}{12}$	$\dfrac{1}{2} - \dfrac{1}{12}t$	1		$\dfrac{1}{12}$	$-\dfrac{1}{3}$
	3	x_3	$\dfrac{5}{4}$	$\dfrac{1}{2} + \dfrac{1}{4}t$		1	$-\dfrac{1}{4}$	
		g	$\dfrac{67}{12}$	$-\dfrac{3}{2} + \dfrac{7}{12}t$	0	0	$-\dfrac{7}{12}$	$-\dfrac{2}{3}$

which for $t = 0$ gives the optimal solution of the problem as it stands.

The parametric solution shown in the tableau is feasible (non-negative) for all values of t, and the simplex criterion is seen to be satisfied as long as the simplex coefficient in the x_1 column is non-positive, i.e., for

$$-\frac{3}{2} + \frac{7}{12}t \leq 0 \text{ or } t \leq \frac{18}{7}.$$

Since a_{11} will normally be ≥ 0 in a diet problem, this means that the solution is optimal for $-2 \leq t \leq 18/7$, or $0 \leq a_{11} \leq 32/7$.

When t passes the critical value $18/7$, the simplex coefficient of x_1 becomes positive. Proceeding by the simplex method we get to Tableau II where x_1 has replaced x_3 in the basis:

			0	4	2	3	0	0
			x_0	x_1	x_2	x_3	x'_1	x'_2
II	2	x_2	$\dfrac{4t-2}{3t+6}$		1	$\dfrac{t-6}{3t+6}$	$\dfrac{2}{3t+6}$	$-\dfrac{1}{3}$
	4	x_1	$\dfrac{5}{t+2}$	1		$\dfrac{4}{t+2}$	$-\dfrac{1}{t+2}$	0
		g	$\dfrac{8t+56}{3t+6}$	0	0	$-\dfrac{7t-18}{3t+6}$	$-\dfrac{8}{3t+6}$	$-\dfrac{2}{3}$

This solution is positive and satisfies the simplex criterion for all $t \geq 18/7$. In this interval the optimal basis consists of x_1 and x_2, but their values and thus also the preference function are now functions of the parameter t; g is a decreasing function of t (i.e., of a_{11}).

The results can be condensed into the following table:

Interval	Tableau	Optimal basis	g	dg/dt $(=dg/da_{11})$
$(-2\leq)t\leq 18/7$, $(0\leq)a_{11}\leq 32/7$	I	x_2, x_3	$\dfrac{67}{12}$	0
$18/7 \leq t$, $32/7 \leq a_{11}$	II	x_1, x_2	$\dfrac{8t+56}{3t+6}$	$\dfrac{-120}{(3t+6)^2}(<0)$

For $t = 18/7$ we get two *alternative optima*: $x_2 = 11/12$, $x_3 = 5/4$ (Tableau I) and $x_1 = 35/32$, $x_2 = 29/48$ (Tableau II), each tableau having a simplex coefficient equal to zero.

Parametric variations in the other a_{ij} can be examined in a similar manner.

3. Next, let us see what happens if one of the right-hand sides is varied parametrically; for example, set $b_1 = 5 + t$. The first simplex tableau then becomes

			0	4	2	3	0	0
			x_0	x_1	x_2	x_3	x_1'	x_2'
I'	2	x_2	$\dfrac{-t+11}{12}$	$\dfrac{1}{2}$	1		$\dfrac{1}{12}$	$-\dfrac{1}{3}$
	3	x_3	$\dfrac{t+5}{4}$	$\dfrac{1}{2}$		1	$-\dfrac{1}{4}$	
		g	$\dfrac{7t+67}{12}$	$-\dfrac{3}{2}$	0	0	$-\dfrac{7}{12}$	$-\dfrac{2}{3}$

The solution is seen to be feasible for $(-5\leq) t \leq 11$, i.e., for $0 \leq b_1 \leq 16$. The simplex criterion is satisfied for all values of t; parametric variation of a right-hand side coefficient affects only the P_0 column, i.e., the values of the basic variables and the preference function.

Hence, for $t > 11$ the simplex coefficients will still have the proper sign, while x_2 becomes negative; in other words, we now have a solution which is *dual feasible* so that the next tableau can be determined by the *dual simplex method*. The variable to replace the negative variable x_2 in the basis is designated by the smallest ratio of a simplex coefficient to the corresponding number in the x_2 row (provided that the latter number is negative); in the present case, x_2' is the only candidate. With x_2' replacing x_2, the familiar transformation rules provide the next tableau, II':

		x_0	4 x_1	2 x_2	3 x_3	0 x'_1	0 x'_2
II'	0 x'_2	$\dfrac{t-11}{4}$	$-\dfrac{3}{2}$	-3		$-\dfrac{1}{4}$	1
	3 x_3	$\dfrac{t+5}{4}$	$\dfrac{1}{2}$		1	$-\dfrac{1}{4}$	
	g	$\dfrac{3t+15}{4}$	$-\dfrac{5}{2}$	-2	0	$-\dfrac{3}{4}$	0

Tableau II' is feasible and optimal for all $t \geq 11$.
The results of the analysis are shown in the following table:

Interval	Tableau	Optimal basis	g	dg/dt $(=dg/db_1)$
$-5 \leq t \leq 11$, $0 \leq b_1 \leq 16$	I'	x_2, x_3	$\dfrac{7t+67}{12}$	$7/12$ (>0)
$11 \leq t$, $16 \leq b_1$	II'	x_3, x'_2	$\dfrac{3t+15}{4}$	$3/4$ (>0)

For $t = 11$ the two tableaux yield the *identical degenerate solutions* $x_2 = 0$, $x_3 = 4$ and $x'_2 = 0$, $x_3 = 4$ because the critical value implies that a basic variable becomes zero.

The preference function is seen to be a piecewise linear function of t. The derivative $dg/dt = dg/db_1$ — discontinuous at the critical value of t — will have a meaningful interpretation according to the nature of the programming problem. If the present problem is interpreted as a diet problem, where b_1 represents, e.g., the minimum amount of vitamin A required in the diet, a unit increase of b_1 from 5 to 6 will increase total cost by 7/12, which is therefore the marginal cost of "producing" a unit more of vitamin A. The number 7/12 recurs (with opposite sign) as the simplex coefficient of x'_1 in Tableau I'; increasing x'_1 by a unit corresponds to decreasing b_1.[1]

In a capacity utilization problem[2], the derivative of the preference function with respect to an additive right-hand side parameter can be interpreted as the increase in total profit caused by a unit increase of the capacity in question, i.e., as the shadow price of the services of the capacity factor.

4. Finally, let a coefficient in the preference function be subjected to parametric variation, for example, $c_1 = 4 + t$.

---
[1] The corresponding dual variable (y_1) has the same interpretation.
[2] See Ch. IV, B.

Following a similar procedure to what we did above, we get the following sequence of basic solutions for decreasing values of the parameter t (the first basis now being optimal for values of t *greater* than a certain critical value):

| | | | 0 | $4+t$ | 2 | 3 | 0 | 0 |
			x_0	x_1	x_2	x_3	x'_1	x'_2
	2	x_2	$\frac{11}{12}$	$\frac{1}{2}$	1		$\frac{1}{12}$	$-\frac{1}{3}$
I″	3	x_3	$\frac{5}{4}$	$\frac{1}{2}$		1	$-\frac{1}{4}$	
		g	$\frac{67}{12}$	$-\frac{2t+3}{2}$	0	0	$-\frac{7}{12}$	$-\frac{2}{3}$
	$4+t$	x_1	$\frac{11}{6}$	1	2		$\frac{1}{6}$	$-\frac{2}{3}$
II″	3	x_3	$\frac{1}{3}$		-1	1	$-\frac{1}{3}$	$\frac{1}{3}$
		g	$\frac{11t+50}{6}$	0	$2t+3$	0	$\frac{t-2}{6}$	$-\frac{2t+5}{3}$
	$4+t$	x_1	$\frac{5}{2}$	1	2		$-\frac{1}{2}$	
III″	0	x'_2	1		-3	3	-1	1
		g	$\frac{5t+20}{2}$	0	-2	$2t+5$	$-\frac{t+4}{2}$	0

The critical values of t are now $-3/2$ in Tableau I″ and $-5/2$ in Tableau II″ (i.e., $c_1 = 5/2$ and $3/2$). For each of these critical prices we get *alternate optimal solutions* because one of the simplex coefficients becomes zero[1].

The complete result can be condensed as follows:

Interval	Tableau	Optimal basis	g	dg/dt $(=dg/dc_1)$
$-4 \leq t \leq -5/2$, $0 \leq c_1 \leq 3/2$	III″	x_1, x'_2	$\frac{5t+20}{2}$	$\frac{5}{2}(>0)$
$-5/2 \leq t \leq -3/2$, $3/2 \leq c_1 \leq 5/2$	II″	x_1, x_3	$\frac{11t+50}{6}$	$\frac{11}{6}(>0)$
$-3/2 \leq t$, $5/2 \leq c_1$	I″	x_2, x_3	$\frac{67}{12}$	0

---

[1] *Exercise:* Show this in detail.

Interpreting the problem as a diet problem, the cost of the optimal diet will increase piecewise linearly with c_1 up to the critical value 5/2, where x_1 leaves the basis so that its price no longer affects the cost of the diet.

In practical applications of linear programming, the c_j will typically be the least stable of the coefficients because they represent, or are directly determined by, market prices which are subject to frequent variation. It is therefore particularly important to carry out sensitivity analyses or compute parametric programmes with the c_j as parameters. The computations are quite easy since, as we have seen, the parameter will in this case occur only in the last row of the simplex tableaux (i.e., in the simplex coefficients), a fact which also makes a multiparametric analysis comparatively simple[1,2].

---

1 Cf. the ice cream example above.

2 *Exercise:* Carry out a parametric programming analysis of the problem

$$x_1 + x_2 \leq 5$$
$$x_1 + 2x_2 \leq 8$$
$$5x_1 + 6x_2 = f = \text{maximum},$$

replacing in turn a_{11}, b_1, and c_1 by a parameter t.

VIII. Integer Linear Programming

A. Integer Programming and Solution by Rounding

1. In some linear programming problems it is required of the optimal solution that the variables, or some of them, should be non-negative integers (0, 1, 2, ...), the interpretation of the problem being such that fractional values would be meaningless or irrelevant. For example, a commodity may be produced only in multiples of one ton, or it has to be shipped in large indivisible units. Problems of this type are dealt with by *integer* (or *discrete*) *linear programming*.

If all of the variables are required to have integral values we have a case of *pure* integer (all-integer) programming; *mixed* integer programming is the case where only a specified set of the variables must be integers, the others being continuous non-negative variables. As a special case of integer programming, some or all of the variables may be required to be either 0 or 1 ("*0−1 programming*").

In a transportation problem where the row and column totals are integers, the optimal solution which we find by the transportation method will automatically satisfy the requirement of integrality, but apart from this special case we cannot be sure of getting an integral solution, so some special procedure is clearly called for.

2. An obvious approach is to solve the problem by the simplex method without paying any attention to integrality, and then *round off* the solution to the nearest feasible solution (unless it happens to be integral, in which case we have already solved the problem). Consider the following problem:

$$2x_1 + 3x_2 \le 14.5$$
$$4x_1 + 1x_2 \le 16.5$$
$$3x_1 + 2x_2 = f = \text{maximum}$$

where the structural variables must have integral values. By the simplex method, requiring only that the variables should be non-negative, we get the fractional solution

$$x_1 = 3.5, \ x_2 = 2.5; \ f = 15.5.$$

There are four possible ways of rounding this solution up or down to the nearest integers,

x_1	4	4	3	3
x_2	3	2	3	2

of which the first is seen to violate both restrictions while the next two violate one restriction each. The fourth solution only is feasible,

$$x_1 = 3, \ x_2 = 2; \ f = 13.$$

This is in fact the best integral solution, as shown geometrically in Fig. 11 where the feasible integral solutions form a "lattice"[1].

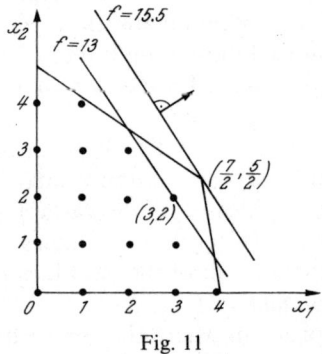

Fig. 11

3. However, quite apart from the fact that the number of such "nearest rounded" solutions to be examined increases fast with the number of variables, we cannot be sure that one of them represents the optimal integral solution.

Consider, for example, the problem

$$\begin{aligned} 2x_1 + 3x_2 &\leq 14 \\ 2x_1 + 1x_2 &\leq 9 \\ 3x_1 + 2x_2 &= f = \text{maximum} \end{aligned} \tag{1}$$

where all of the variables — including the slack variables, x_3 and x_4 — must have integral values (0, 1, 2, ...). Solving by the simplex method without taking integrality into account (but of course with non-negativity requirements), we have the following simplex tableaux:

---

1 *Exercise:* Indicate what Fig. 11 would like like if only x_1 was required to be integral, and find the optimal solution.

Solution by Cuts. Pure Case

			0 x_0	3 x_1	2 x_2	0 x_3	0 x_4
I	0	x_3	14	2	3	1	0
	0	x_4	9	2	1	0	1
		f	0	-3	-2	0	0
II	0	x_3	5	0	2	1	-1
	3	x_1	9/2	1	1/2	0	1/2
		f	27/2	0	$-1/2$	0	3/2
III	2	x_2	5/2	0	1	1/2	$-1/2$
	3	x_1	13/4	1	0	$-1/4$	3/4
		f	59/4	0	0	1/4	5/4

The optimal non-negative solution is that of Tableau III,

$$x_1 = 13/4,\ x_2 = 5/2,\ x_3 = 0,\ x_4 = 0;\ f = 59/4 \qquad (2)$$

which is fractional in x_1 and x_2. As Fig. 12 shows, the best integral solution is

$$x_1 = 4,\ x_2 = 1,\ x_3 = 3,\ x_4 = 0;\ f = 14 \qquad (3)$$

which cannot be derived from (2) by rounding off to nearest integers.

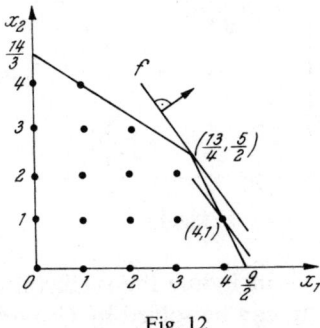

Fig. 12

In other words, solution by rounding will not in general lead to the best integral solution.

B. Solution by Cuts. Pure Case

1. Another approach is to reduce the set of feasible solutions by introducing *additional restrictions* ("*cuts*" or "*cutting planes*") such that a subset containing the non-integral optimum is cut off, without removing any integer feasible solutions. Applying suitable cuts we end up with a

reduced feasible region which has the best integral solution as one of its corners; this solution then comes out as the optimal solution if the simplex method is applied to the problem with the cuts added to the side conditions.

In the example above — cf. (1) and Fig. 12 — it is clear that the cuts

$$x_1 + x_2 \leq 5 \quad (4)$$

$$x_1 \quad \leq 4 \quad (5)$$

$$x_2 \leq 4 \quad (6)$$

will do the trick, as shown in Fig. 13. (4)—(6) reduce the set of feasible solutions to the "convex hull" of integral feasible solutions to (1), and the best integral solution will clearly be one of the corners of this reduced region — in this case, (3). However, a glance at Fig. 13 will show that (4) alone is sufficient. Applying the simplex method to (1) with (4) added to the side conditions will automatically produce the optimal integer solution[1]. There is no point in introducing more additional constraints than we need. The question is how to generate a minimum of cuts which will solve the problem.

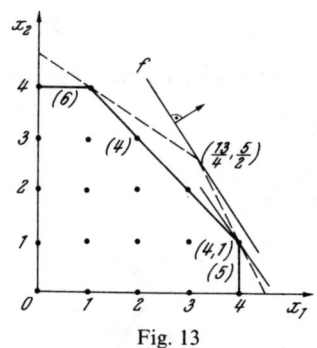

Fig. 13

2. When *all* variables — including the slack variables — are required to be integers, the problem can be solved by Gomory's *algorithm for pure integer programming*[2]. This method, which we shall now demonstrate on the example (1), is based on successive generation of additional constraints (cuts).

The first step is to *solve the problem without taking the integrality requirements into account*. This we have already done in Tableaux I—III above.

---

[1] *Exercise:* Show this.

[2] Cf. S. Vajda (1961), Ch. 10 and (1962), Ch. XXIV. — For a general description see Appendix, E, below.

Next, in the optimal tableau (III) *select a basic variable which is fractional in the solution*, say x_2 ($=5/2$). From the row of x_2 we then construct an additional side condition where the coefficients are determined as follows.

Each coefficient in the row of x_2 is split into the *largest lower integer* (not necessarily positive) and a *proper fraction* (or zero if the coefficient is integer):

	x_0	x_1	x_2	x_3	x_4
x_2	$5/2$	0	1	$1/2$	$-1/2$
	$=2+\frac{1}{2}$	$=0+0$	$=1+0$	$=0+\frac{1}{2}$	$=-1+\frac{1}{2}$

A *new row* containing these fractions with opposite signs is added to the tableau, at the same time adding a *new column* representing an auxiliary variable, x_5, which is to serve as the extra basic variable required at this step because we now have one restriction more; hence the coefficients in the new column are set $=0$ except that the coefficient in the row of x_5 is set $=1$. This leads to *Tableau III a*:

				0	3	2	0	0	0
				x_0	x_1	x_2	x_3	x_4	x_5
	2	x_2		$5/2=2+\frac{1}{2}$	0	1	$1/2$	$-1/2$	0
IIIa	3	x_1		$13/4$	1	0	$-1/4$	$3/4$	0
	0	x_5		$-1/2$	0	0	$-1/2$	$-1/2$	1
		f		$59/4$	0	0	$1/4$	$5/4$	0
	2	x_2		2	0	1	0	-1	1
IV	3	x_1		$7/2$	1	0	0	1	$-1/2$
	0	x_3		1	0	0	1	1	-2
		f		$29/2$	0	0	0	1	$1/2$

The new row merely defines the auxiliary variable x_5,

$$x_5 = -\frac{1}{2} + \frac{1}{2} x_3 + \frac{1}{2} x_4;$$

it imposes a restriction on the original variables only when combined with the non-negativity requirement $x_5 \geq 0$, which is seen to imply the constraint

$$\frac{1}{2} x_3 + \frac{1}{2} x_4 \geq \frac{1}{2}. \tag{7}$$

As we shall see later, this inequality cuts off part of the original set of feasible solutions (including the solution in Tableau III) without excluding any feasible integral solutions; and x_5 will be an integer when the other variables are integral.

Tableau IIIa gives a basic solution in three variables, corresponding to the number of side conditions. The simplex coefficients have the proper sign required for an optimum, but the basic solution is negative in x_5. Hence we must apply the *dual simplex method* to get to a non-negative basic solution. x_5 is the variable to be replaced; the new basic variable is designated by

$$\min\left(\frac{1/4}{1/2}, \frac{5/4}{1/2}\right) = \frac{1/4}{1/2}$$

so that x_5 is replaced by x_3. This leads to *Tableau IV* above.

x_2 has now become an integer, but x_1 is still fractional ($=7/2$), so the procedure is repeated: we introduce another auxiliary variable x_6 (required to be ≥ 0 and integral) and a corresponding row and column; the coefficients in the new row are determined by splitting the coefficients in the row of x_1 in the manner described above, changing the signs of the fractions. In this way we get *Tableau IVa*.

			0 x_0	3 x_1	2 x_2	0 x_3	0 x_4	0 x_5	0 x_6
IVa	2	x_2	2	0	1	0	-1	1	0
	3	x_1	$7/2 = 3 + \frac{1}{2}$	1	0	0	1	$-1/2$	0
	0	x_3	1	0	0	1	1	-2	0
	0	x_6	$-1/2$	0	0	0	0	$-1/2$	1
		f	$29/2$	0	0	0	1	$1/2$	0
V	2	x_2	1	0	1	0	-1	0	2
	3	x_1	4	1	0	0	1	0	-1
	0	x_3	3	0	0	1	1	0	-4
	0	x_5	1	0	0	0	0	1	-2
		f	14	0	0	0	1	0	1

The new row in Tableau IVa, together with $x_6 \geq 0$, implies a further restriction

$$x_6 = -\frac{1}{2} + \frac{1}{2}x_5 \geq 0 \quad \text{or} \quad \frac{1}{2}x_5 \geq \frac{1}{2}, \tag{8}$$

which can be shown to represent another cut.

Applying the dual simplex method we proceed to *Tableau V* which represents an integral basic solution in all variables; the simplex coefficients all have the proper sign. This completes the calculations; the solution corresponds to (3), the optimal integral solution[1]

3. The additional restrictions which we have introduced, (7) and (8), cut off parts of the original set of feasible solution, as can be shown in an (x_1, x_2) diagram as follows.

Solving the original side conditions of (1) for x_3 and x_4 and substituting in (7), we have

$$x_1 + x_2 \leq \frac{11}{2}. \tag{7a}$$

Similarly, (8) can be expressed as a restriction on x_1 and x_2; inserting the definition of x_5

$$x_5 = -\frac{1}{2} + \frac{1}{2} x_3 + \frac{1}{2} x_4$$

and expressing x_3 and x_4 in terms of x_1 and x_2 as above we get

$$x_1 + x_2 \leq 5. \tag{8a}$$

Hence the problem we have solved in Tableau V is the original problem (1) with the additional side conditions (7a)—(8a), of which (7a) turns out to be redundant. The cuts which they represent are indicated by shading in Fig. 14.

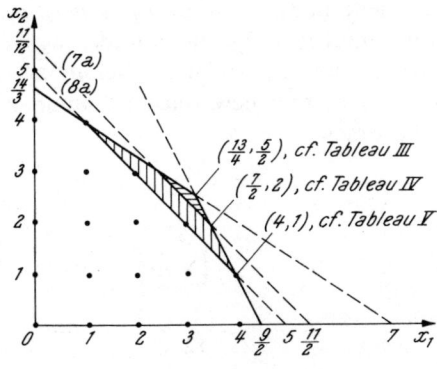

Fig. 14

C. Solution by Cuts. Mixed Case

Consider again problem (1), but assume now that *only* x_1 is required to be integral. All variables must be non-negative. Then we have an example of *mixed* integer programming.

---

[1] *Exercise:* Solve the problem (1) by Gomory's algorithm, generating a further restriction in Tableau III *from the row of* x_1.

Problems of this type can be dealt with by a related method, *Gomory's algorithm for mixed integer programming*[1]. Again, the starting point is the optimal non-integral solution (Tableau III above), and the procedure consists in introducing successive cuts and corresponding auxiliary variables which must be non-negative. The additional restrictions (cuts), however, are generated in a somewhat different way.

In Tableau III, the row of x_1 is

	x_0	x_1	x_2	x_3	x_4
x_1	$13/4 = 3 + \frac{1}{4}$	1	0	$-1/4$	$3/4$

As in the pure case, the value of x_1 is split into the largest lower integer and a non-negative fraction,

$$\bar{x}_1 = N_1 + f_1 \text{ or } \frac{13}{4} = 3 + \frac{1}{4}.$$

In the new row to be introduced — corresponding to an auxiliary variable x_5 — write $-f_1 = -1/4$ in the x_0 column, as we did before. The other coefficients in the new row are determined from those of the x_1 row as follows. If a coefficient in the row of x_1 is *positive*, multiply it by (-1); if it is *negative*, multiply by the number $g_1 = f_1/(1-f_1) = (1/4)/(3/4) = 1/3$. Finally, write 1 for the coefficient belonging to the column of x_5. All other coefficients in this new (basic) column are set $= 0$. The additional row then becomes

	x_0	x_1	x_2	x_3	x_4	x_5
x_5	$-\frac{1}{4}$	0	0	$\frac{1}{3} \cdot \left(-\frac{1}{4}\right) = -\frac{1}{12}$	$(-1) \cdot \frac{3}{4} = -\frac{3}{4}$	1

The requirement $x_5 \geq 0$ implies the restriction

$$x_5 = -\frac{1}{4} + \frac{1}{12} x_3 + \frac{3}{4} x_4 \geq 0;$$

substituting the expressions for x_3 and x_4 which we get from the original side conditions this is seen to imply

---

[1] Cf. S. Vajda (1962), Ch. XXIV. — A general description of the algorithm is given below in the Appendix, E.

$$5 x_1 + 3 x_2 \leq 23, \tag{9}$$

which cuts off part of the set of feasible solutions to (1)[1].

Adding the new row and the new column to Tableau III we get *Tableau III b*, from which we proceed by the dual simplex method to *Tableau IV b*; the latter gives the solution

$$x_1 = 3, \ x_2 = 8/3, \ x_3 = 0, \ x_4 = 1/3; \ f = 43/3.$$

			0	3	2	0	0	0
			x_0	x_1	x_2	x_3	x_4	x_5
IIIb	2	x_2	5/2	0	1	1/2	−1/2	0
$f_1 = 1/4$	3	x_1	$13/4 = 3 + \frac{1}{4}$	1	0	−1/4	3/4	0
$g_1 = 1/3$	0	x_5	−1/4	0	0	−1/12	−3/4	1
		f	59/4	0	0	1/4	5/4	0
IVb	2	x_2	8/3	0	1	5/9	0	−2/3
	3	x_1	3	1	0	−1/3	0	1
	0	x_4	1/3	0	0	1/9	1	−4/3
		f	43/3	0	0	1/9	0	5/3

The solution is integral in x_1, so we have solved the problem. The solution for x_1 and x_2 is seen to satisfy (9) and the first side condition of (1) in equality form[2, 3].

D. Solution by a Branch-and-Bound Procedure

Another approach, applicable to pure as well as mixed cases, is to use a *branch-and-bound* technique, a general method which has been successfully applied to a wide variety of problems in operations research. We shall now show how problem (1) can be solved by such a procedure, assuming that all variables are required to be ≥ 0 and integral. (It would make no difference in this case if only the structural variables x_1 and x_2 were required to be integers).

Again, the starting point is the optimal non-integral solution, here called *solution 0*, as shown in *Tableau 0* (= Tableau III above).

---

1 *Exercise:* Show that the optimal non-integral solution is cut off by (9).

2 *Exercise:* Illustrate the cut and the solution geometrically (cf. Fig. 12).

3 *Exercise:* Solve problem (1) by the same algorithm when x_2, instead of x_1, is required to be integral. Illustrate geometrically.

		x_0	x_1	x_2	x_3	x_4
	x_1	$\frac{13}{4}=3\frac{1}{4}$	1	0	$-1/4$	$3/4$
0	x_2	$\frac{5}{2}=2\frac{1}{2}$	0	1	$1/2$	$-1/2$
	f	$\frac{59}{4}=14\frac{3}{4}$	0	0	$1/4$	$5/4$

Consider an arbitrary basic variable which must be integral but is fractional in the solution, say x_1. Its value $(3^1/_4)$ is between 3 and 4. Then *separate the set of feasible solutions into disjoint subsets*, each defined by (1) with a further restriction:

$$x_1 \leq 3 \quad \text{or} \quad x_1 + x_5 = 3 \quad (10\text{a})$$
and
$$x_1 \geq 4 \quad \text{or} \quad x_1 - x_5 = 4 \quad (10\text{b})$$

where x_5 is a non-negative auxiliary variable. The optimal integral solution must be in one of these subsets, i.e., it must be a feasible solution to *either* (1) with (10a) *or* (1) with (10b). The rest of the original feasible region, as defined by (1) with $3 < x_1 < 4$, can be disregarded because x_1 must be an integer.

The next step is to solve the two restricted problems. Introducing (10a) we get *solution* 1; with (10b) we get *solution* 2. These logical alternatives can be illustrated by a *branching* from solution 0, cf. Fig. 15:

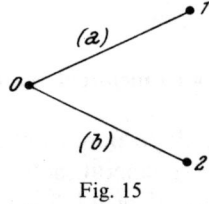

Fig. 15

(10a) implies the further restriction

$$x_5 = 3 - x_1 \geq 0;$$

expressing x_1 in terms of the non-basic variables in Tableau 0,

$$x_1 = \frac{13}{4} + \frac{1}{4} x_3 - \frac{3}{4} x_4,$$

we get

$$x_5 = -\frac{1}{4} - \frac{1}{4} x_3 + \frac{3}{4} x_4 \quad (11\text{a})$$

which, with $x_5 \geq 0$, represents the cut defined by (10a). Introducing this further restriction in Tableau 0, with x_5 as an additional basic variable, we get *Tableau 0a*, from which the dual simplex method leads to *Tableau 1*. The latter gives *solution 1*.

		x_0	x_1	x_2	x_3	x_4	x_5
	x_1	$\frac{13}{4} = 3\frac{1}{4}$	1	0	$-1/4$	$3/4$	0
0a	x_2	$\frac{5}{2}$	0	1	$1/2$	$-1/2$	0
	x_5	$-\frac{1}{4}$	0	0	$1/4$	$-3/4$	1
	f	$\frac{59}{4} = 14\frac{3}{4}$	0	0	$1/4$	$5/4$	0
	x_1	3	1	0	0	0	1
1	x_2	$\frac{8}{3} = 2\frac{2}{3}$	0	1	$1/3$	0	$-2/3$
	x_4	$\frac{1}{3}$	0	0	$-1/3$	1	$-4/3$
	f	$\frac{43}{3} = 14\frac{1}{3}$	0	0	$2/3$	0	$5/3$

Similarly, (10b) can be shown to imply

$$x_5 = -\frac{3}{4} + \frac{1}{4}x_3 - \frac{3}{4}x_4 \quad (\geq 0). \tag{11b}$$

Introducing this cut in Tableau 0 leads to *Tableau 0b*, from which we proceed to *Tableau 2* with *solution 2*, the optimal solution to (1) with (10b).

		x_0	x_1	x_2	x_3	x_4	x_5
	x_1	$\frac{13}{4} = 3\frac{1}{4}$	1	0	$-1/4$	$3/4$	0
0b	x_2	$\frac{5}{2}$	0	1	$1/2$	$-1/2$	0
	x_5	$-\frac{3}{4}$	0	0	$-1/4$	$3/4$	1
	f	$\frac{59}{4} = 14\frac{3}{4}$	0	0	$1/4$	$5/4$	0
	x_1	4	1	0	0	0	-1
2	x_2	1	0	1	0	1	2
	x_3	3	0	0	1	-3	-4
	f	14	0	0	0	2	1

The results of the branching are indicated in Fig. 16, where the value of f in solution no. h is denoted by I_h if it is integral in those variables which are required to be so, and by F_h if it is fractional in one or more such variables.

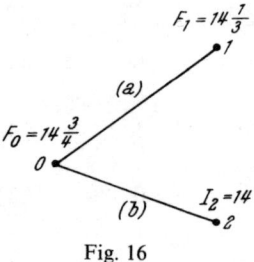

Fig. 16

Passing from left to right in Fig. 16, the value of f will be non-decreasing since a better solution can never be attained by cutting off part of the feasible region. Hence F_0 is an *upper bound* on F_1 and I_2, and these are in turn upper bounds on the values of f which we get by further branchings from solutions 1 and 2. F_1 and I_2 are in fact smaller than F_0. Now the question is whether further branching can improve the solution.

It is obvious that nothing is to be gained by branching from *solution 2* because it is already *integral* and further branching can only lead to inferior solutions. Whether it pays to branch from *solution 1*, which is *fractional* in two of the variables, depends on whether F_1 is greater or smaller than I_2. F_1 is an upper bound on f in branchings from solution 1. Hence, if $F_1 \leq I_2$ we would have $f \leq F_1 \leq I_2$ for all such branching solutions and there would be no point in examining them. In the example, however, we have $F_1 > I_2$, so it is conceivable that there exists a solution in which f is less than the upper bound $F_1 (= 14^1/_3)$ but greater than $I_2 (= 14)$.

We therefore *repeat the branching operation*, now starting from solution 1. Consider, for example, $x_2 (= 2^2/_3)$, which has a value between 2 and 3. Then the following two alternative restrictions are introduced:

$$x_2 \leq 2 \quad \text{or} \quad x_2 + x_6 = 2 \tag{12a}$$

$$x_2 \geq 3 \quad \text{or} \quad x_2 - x_6 = 3 \tag{12b}$$

where x_6 is a non-negative variable; expressing x_2 in terms of the non-basic variables (cf. the row of x_2 in Tableau 1),

$$x_2 = \frac{8}{3} - \frac{1}{3} x_3 + \frac{2}{3} x_5,$$

and substituting in (12a) and (12b), we get the extra row in *Tableaux 1a and 1b*. Proceeding by the dual simplex method we get to *Tableau 3* and *Tableau 4* respectively:

Solution by a Branch-and-Bound Procedure

		x_0	x_1	x_2	x_3	x_4	x_5	x_6
	x_1	3	1	0	0	0	1	0
	x_2	$\frac{8}{3}=2\frac{2}{3}$	0	1	1/3	0	$-2/3$	0
1a	x_4	$\frac{1}{3}$	0	0	$-1/3$	1	$-4/3$	0
	x_6	$-\frac{2}{3}$	0	0	$-1/3$	0	2/3	1
	f	$\frac{43}{3}=14\frac{1}{3}$	0	0	2/3	0	5/3	0
	x_1	3	1	0	0	0	1	0
	x_2	2	0	1	0	0	0	1
3	x_4	1	0	0	0	1	-2	-1
	x_3	2	0	0	1	0	-2	-3
	f	13	0	0	0	0	3	2

		x_0	x_1	x_2	x_3	x_4	x_5	x_6
	x_1	3	1	0	0	0	1	0
	x_2	$\frac{8}{3}=2\frac{2}{3}$	0	1	1/3	0	$-2/3$	0
1b	x_4	$\frac{1}{3}$	0	0	$-1/3$	1	$-4/3$	0
	x_6	$-\frac{1}{3}$	0	0	1/3	0	$-2/3$	1
	f	$\frac{43}{3}=14\frac{1}{3}$	0	0	2/3	0	5/3	0
	x_1	$\frac{5}{2}=2\frac{1}{2}$	1	0	1/2	0	0	3/2
	x_2	3	0	1	0	0	0	-1
	x_4	1	0	0	-1	1	0	-2
4	x_5	$\frac{1}{2}$	0	0	$-1/2$	0	1	$-3/2$
	f	$\frac{27}{2}=13\frac{1}{2}$	0	0	3/2	0	0	5/2

The results of this branching are shown in Fig. 17[1].

---

[1] A diagram of this type is known as a *tree*. In a branch-and-bound tree the nodes represent solutions and the branches represent separations of the feasible region.

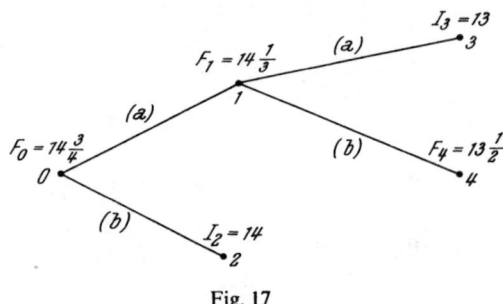

Fig. 17

Fig. 17 shows that further branching from solution 4 might conceivably give an integral solution with $f > I_3 = 13$. However, since $F_4 = 13^{1}/_{2}$ is now an upper bound on f and since we have already found an integral solution which is better — viz. $I_2 = 14$ — it would be a waste of time to explore these branchings. The best integral solution is evidently solution 2, i.e.,

$$x_1 = 4, \ x_2 = 1, \ x_3 = 3, \ x_4 = 0; \ f = 14,$$

which corresponds to what we found above by other methods[1].

E. 0-1 Programming

1. In some linear programming problems the variables (or a specified subset of variables) are required to have the values 0 or 1 only. Such *0 − 1 variables*, or *binary* variables, may be introduced to represent combinatorial elements (either-or) within the framework of a linear model.

Requirements of this type,

$$x_j = 0 \text{ or } 1, \qquad (13)$$

can be handled by general integer programming methods, replacing (13) by the restrictions

$$\begin{aligned} x_j &\leq 1 \\ x_j &\geq 0 \text{ and integral} \end{aligned} \qquad (13a)$$

which together exclude all values but 0 and 1. A simpler but often dangerous procedure is to replace (13) by

$$0 \leq x_j \leq 1, \qquad (13b)$$

solve by the standard simplex method and subsequently round off the solution to 0 or 1 — taking care that the rounded solution satisfies the original side conditions.

---

[1] *Exercise*: Solve (1) by the branch-and-bound procedure, now basing the separation on the row of x_2 in Tableau 0.

2. Problems of this nature occur in *linear investment planning*, where a decision variable x_j is associated with each investment project. The fact that a project can be either accepted or rejected is then expressed by $x_j=1$ or $x_j=0$ respectively [1].

Let Y_1 and Y_2 be the net present values of two *alternative* (*mutually exclusive*) projects. The choice between them can then be expressed in the model

$$Y_1 x_1 + Y_2 x_2 = Y = \max$$
$$x_1 + x_2 \leq 1$$
$$x_j = 0 \text{ or } 1 \quad (j=1, 2).$$

The advantage of this formulation is that combinatorial solution is avoided. With a large number of projects it soon becomes impracticable to enumerate and compare all possible combinations of projects (including the zero investment).

With a view to solution by integer programming methods the model can be written

$$Y_1 x_1 + Y_2 x_2 = Y = \max$$
$$x_1 + x_2 \leq 1$$
$$x_j \geq 0 \text{ and integral } (j=1, 2);$$

there is no need in this case to introduce the restrictions $x_j \leq 1$ $(j=1, 2)$ since the non-negativity requirements and the restriction $x_1 + x_2 \leq 1$ exclude values greater than 1.

If project 1 is a *prerequisite* of project 2, Y is to be maximized subject to

$$x_2 \leq x_1, \quad x_j = 0 \text{ or } 1 \quad (j=1, 2),$$

which may be rewritten

$$x_2 \leq x_1 \leq 1, \quad x_j \geq 0 \text{ and integral}$$

so that methods of integer programming can be applied.

3. A similar application of $0-1$ programming is made in the so-called *fixed-cost problem*. A simple example is the following.

A company producing x units per period of a commodity wishes to maximize profit per period,

$$f = px - c(x),$$

where p is the selling price per unit and $c(x)$ is cost per period, subject to a capacity restriction $x \leq b$. The cost function has the form

$$c = \begin{cases} ax+k & \text{for } x > 0 \\ 0 & \text{for } x = 0 \end{cases} \tag{14}$$

---

[1] Cf. Ch. IV, E.

where a is variable cost per unit whereas k is a fixed cost element incurred only when the activity level x is positive; k may represent, for example, a *setup cost*.

In order to avoid a combinatorial procedure with two alternative cost functions (or, in the case of several activities, a larger number of combinations of positive activity levels and zeroes) the cost function may be written

$$c = ax + k\delta \qquad (15)$$

where

$$\delta = \begin{cases} 1 & \text{for } x > 0 \\ 0 & \text{for } x = 0; \end{cases} \qquad (16)$$

to ensure that (16) is satisfied, introduce an additional restriction

$$x \leq b\delta \qquad (17)$$

where b is an upper bound of x (in this case, the capacity limit is used), and the $0-1$ requirement

$$\delta = 0 \text{ or } 1 \qquad (18)$$

or, with a view to solution by integer programming,

$$0 \leq \delta \leq 1 \text{ and } \delta \text{ integral}[1]. \qquad (18a)$$

Then the problem becomes a case of mixed integer linear programming,

$$f = px - ax - k\delta = \text{maximum}$$

subject to
$$x \leq b\delta$$
$$0 \leq \delta \leq 1$$
$$x \geq 0, \ \delta \text{ integral}$$

(the capacity restriction $x \leq b$ being now redundant, cf. (17)—(18)). For $x > 0$, (17) precludes δ from being zero, hence it must be $=1$; for $x = 0$, the maximization of f will see to it that $\delta = 0$.

Let us apply this technique to the blending problem in Ch. IV, A, pp. 30—32.

$$\begin{aligned} 2x_1 + 1x_2 &\leq 10 \\ 1x_1 + 1x_2 &\leq 6 \\ 2x_1 + 3x_2 &\leq 15 \\ 3x_1 + 2x_2 &= f = \max, \end{aligned} \qquad (19)$$

where x_1 and x_2 are two finished products (grades of gasoline) produced by blending the same three raw materials (crude gasolines) in different

---

[1] Cf. G. B. Dantzig (1963), Ch. 26.

proportions. Now assume that there is a fixed cost of $k_1 = 0.4$ (\$/period) associated with the blending of the first product if — and only if — it is actually produced, and similarly that the blending of crude gasolines into the second product costs $k_2 = 0.6$ \$/period for any positive quantity made. Then the preference function becomes

$$f = 3 x_1 + 2 x_2 - 0.4 \delta_1 - 0.6 \delta_2 = \max,$$

and to the side conditions — which are seen to impose an upper bound of 5 on both x_1 and x_2 — we must add the further restrictions

$$x_1 \leq 5 \delta_1$$
$$x_2 \leq 5 \delta_2$$

and $\delta_j = 0$ or 1, or

$$0 \leq \delta_j \leq 1, \quad \delta \text{ integral } (j = 1, 2).$$

This trick makes it possible to solve the problem by mixed integer programming[1]. Otherwise it would have to be solved combinatorially, that is, separately for each of the four possible combinations of positive and zero values of the x_j.

This example may also be interpreted as a problem of optimal capacity utilization where both products are processed on three machines. The fixed costs k_j will in this case represent setup costs of the respective products.

A similar procedure is applicable when *setup times* come into the picture. Let (19) be a problem of capacity utilization where the right-hand sides of the restrictions represent machine hours available per period, and let t_{ij} denote setup time for product no. j on machine no. i. Then the right-hand sides, adjusted for setup times, become

$$10 - t_{11} \delta_{11} - t_{12} \delta_{12}$$
$$6 - t_{21} \delta_{21} - t_{22} \delta_{22}$$
$$15 - t_{31} \delta_{31} - t_{32} \delta_{32},$$

and the following additional constraints must be introduced:

$$x_1 \leq 5 \delta_{i1} \quad (i = 1, 2, 3)$$
$$x_2 \leq 5 \delta_{i2} \quad (i = 1, 2, 3)$$
$$\delta_{ij} = 0 \text{ or } 1 \quad (i = 1, 2, 3; j = 1, 2)^2.$$

---

1 *Exercise:* Solve the problem in this form.

2 If the planner has some preconceived notion of which of the x_j will be positive in the optimum, or if setup times are insignificant in proportion to the capacities, he may find it easier to apply the simple ad-hoc procedure sketched on p. 51 above (cf. also p. 30 on setup costs).

F. Computer Solution

In recent years a number of computer algorithms for integer linear programming have been developed for pure and mixed cases as well as for 0−1 programming. Some of them are based on improved variants of Gomory's methods (cuts), but for the most part they appear to be comparatively inefficient because of the large number of simplex iterations required. Algorithms based upon branch-and-bound procedures would in general seem to be more promising[1].

---

[1] For computer algorithms see, e. g., E. M. L. Beale (1968), Ch. 15. — A survey of existing algorithms is found in A. M. Geoffrion and R. E. Marsten (1972).

IX. Decomposition

A. Decomposition and Decentralized Planning

1. The structure of a linear programming problem may be such that it can be *decomposed into independent subproblems* which can be solved separately.
Consider, for example, the following problem

$$3 x_1 + 4 x_2 + 1 x_3 \qquad \leq 10 \qquad (1)$$
$$1 x_1 + 3 x_2 + 2 x_3 \qquad \leq 5 \qquad (2)$$
$$2 x_4 + 3 x_5 \leq 13 \qquad (3)$$
$$1 x_4 + 2 x_5 \leq 8 \qquad (4)$$
$$3 x_1 + 6 x_2 + 2 x_3 + 7 x_4 + 12 x_5 = f = \max,$$

which may be interpreted as a problem of optimal capacity utilization in a firm organized in two divisions. The first division manufactures the three first products on two machines subject to capacity limitations; division no. 2 produces commodities no. 4 and 5, using two other machines with limited capacities.

Since each division has its own group of variables and its own capacity restrictions, the *total* planning problem can be solved by letting each division solve its own *subproblem*, viz.

(i) $f_1 = 3 x_1 + 6 x_2 + 2 x_3 = \max$ subject to (1)—(2), and
(ii) $f_2 = 7 x_4 + 12 x_5 = \max$ subject to (3)—(4);

clearly the solutions to the subproblems,

$$x_1 = 2, \ x_2 = 1, \ x_3 = 0; \ f_1 = 12$$

and

$$x_4 = 2, \ x_5 = 3; \ f_2 = 50,$$

together constitute an optimal solution of the total problem (with $f = f_1 + f_2 = 62$). Hence complete decentralization of the firm's planning is possible and will lead to decisions which fit into the overall optimality criterion.

2. Suppose, however, that in addition to (1)—(4) there is a *common restriction*

$$2x_1 + 3x_2 + 1x_3 + 2x_4 + 2x_5 \leq 12 \tag{5}$$

representing a fifth machine on which all five products must also be processed. The variables of the two divisions are now connected by (5); the subproblems may be said (somewhat loosely) to be "*almost independent*", but if their solutions above are inserted in (5) we find that the common restriction is violated. The subprogrammes together require 17 machine hours on the fifth machine but only 12 hours are available. The production programme will evidently have to be revised.

This can of course be done by solving the total problem with five side conditions. However, an alternative procedure immediately suggests itself: it might be possible somehow to start with the solutions to the subproblems (i) and (ii), revise them, and "sew them together" in such a way as to satisfy the common restriction (5). Such a procedure is called *solution by decomposition*.

It is a condition that the structure of the coefficient matrix is such that the variables can be arranged in two or more groups each having its particular set of restrictions, to which must be added one or more common restrictions connecting the groups. In other words, the matrix must look like that shown in Fig. 18 (easily generalized to cases of more than two groups of variables), though it may take some previous rearranging of rows and columns to reveal such a structure.

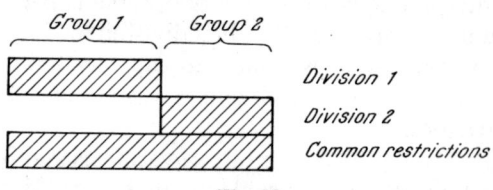

Fig. 18

Structures of this type are typically encountered in linear problems of production or investment planning in large firms organized in divisions (sectors) each having a specific set of restrictions, e.g., capacity limitations or constraints representing investment dependencies.

The *computational* advantages of solution by decomposition are due to the fact that subproblems of smaller dimension are more easily solved; computer time increases fast with the number of restrictions. In addition there may be considerable *organizational* and *psychological* advantages in decentralizing the planning and decision process.

3. A possible procedure for solving the overall problem is to start from the optimal subprogrammes and use the slack variables of the

common restrictions as the "missing" basic variables in the total problem. If the basic solution is negative in one or more slack variables — as it will be when the combined subprogrammes violate a common restriction — the dual simplex method is applied until a non-negative basic solution is attained.

Applying this procedure to the problem above, we first get the following simplex tableau, where x_6—x_{10} are the slack variables of (1)—(5). The first two rows correspond to the optimal solution of subproblem (i) and the next two to the solution of (ii), whereas the fifth row represents the common restriction (5) with the basic variable x_{10} expressed in terms of the non-basic variables of the subproblems. The simplex coefficients are not affected by the introduction of the fifth row because the coefficient of x_{10} in the preference function is zero, hence the dual simplex method is applicable[1].

		x_0	x_1	x_2	x_3	x_4	x_5	x_6	x_7	x_8	x_9	x_{10}
(1)–(2)	x_1	2	1	0	−1			0.6	−0.8			
	x_2	1	0	1	1			−0.2	0.6			
(3)–(4)	x_4	2				1	0			2	−3	
	x_5	3				0	1			−1	2	
(5)	x_{10}	−5	0	0	0	0	0	−0.6	−0.2	−2	2	1
	f	62	0	0	1	0	0	0.6	1.2	2	3	0

After a number of transformations of the tableau it turns out that there are two alternate optimal solutions:

x_1	x_2	x_3	x_4	x_5	x_6	x_7	x_8	x_9	x_{10}	f
0	1	1	0	4	5	0	1	0	0	56
0	4/3	0	0	4	14/3	1	1	0	0	56

In the present case this is not a very efficient procedure; it is only in the fourth tableau that the first optimal solution appears[2].

Alternatively the problem can be solved by the standard (primal) simplex method, introducing an *artificial variable* x_{11} in (5) as the fifth basic variable (with the coefficient $-M$ in the preference function). Instead of $x_{10} = -5$ we get $x_{11} = 5$ in the tableau, from which we proceed by the

---

1 *Exercise:* What would the tableau look like if the right-hand side of (5) had been 18 instead of 12, and what would the optimal solution be?

2 The following changes in the basis are made: x_{10} is first replaced by x_8, then x_4 by x_6, and finally x_1 by x_3.

simplex method until x_{11} has been thrown out of the basis and all simplex coefficients are non-negative [1].

4. A different approach is to apply an *iterative procedure* where the subproblems are successively adjusted in such a way that we get *a sequence of suboptima converging to the total optimum*. This approach, applied as a *decentralization* mechanism, has the advantage of involving the lower level of the hierarchy in the firm's planning and decision making in a more active way.

On the basis of this idea a number of *decomposition algorithms* have been developed in which *shadow prices* are used to coordinate the lower-level plans or decisions such as to satisfy the overall criterion of optimality [2]. We shall demonstrate two such procedures on a simple example, to be interpreted as a problem of optimal capacity utilization:

$$2 x_1 \qquad \leq 8 \qquad (6)$$

$$3 x_2 \leq 9 \qquad (7)$$

$$1 x_1 + 2 x_2 \leq 8 \qquad (8)$$

$$10 x_1 + 10 x_2 = f = \max$$

where the slack variables will be denoted by x_3, x_4, and x_5. The firm is organized in two divisions each of which produces one commodity, cf. Fig. 19, where v_i ($i = 1, 2, 3$) indicates machine hours used per period on the ith machine.

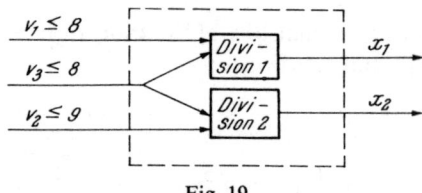

Fig. 19

Provisionally disregarding the common capacity restriction (8), the divisional subproblems can be solved independently. The solutions are

$$x_1 = 4; \quad f_1 = 40$$

and

$$x_2 = 3; \quad f_2 = 30.$$

Inserting them in (8) we have

---

[1] Cf. pp. 81 ff.

[2] The first decomposition algorithm (which will not be dealt with here) was developed by Dantzig and Wolfe, cf. G. B. Dantzig and P. Wolfe (1961); see also W. J. Baumol and T. Fabian (1964).

$$v_3 = 1 \cdot x_1 + 2 \cdot x_2 = 10 > 8,$$

which means that there is not enough capacity on the third machine; the combined subprogrammes do not constitute a feasible solution to the total problem ($x_5 = -2$). This is illustrated geometrically in Fig. 20[1].

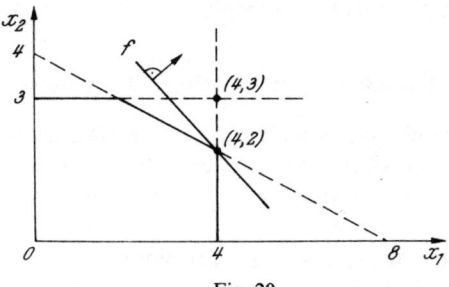

Fig. 20

The optimal total solution is seen to be the point (4, 2), not the combined subprogrammes (4, 3). One of the divisions will have to cut down its production programme; in this case it is division 2 that has to suffer a reduction, but we know this only when the total problem has been solved.

Since the whole problem is due to the fact that the two divisions use too much of the common capacity factor, there are two obvious ways of tackling it: by *direct allocation* of the capacity factor or by *charging a price* for its services.

The first procedure can be elaborated as follows. The central authority (the top level) decides upon a *tentative allocation of the capacity factor* (in this case, $\bar{v}_3 = 8$ capacity units) *to the divisions*. Then *the divisions solve the respective subproblems* corresponding to this allocation. The combined subprogrammes constitute a feasible solution to the total problem; the solution represents a total optimum if and only if *the shadow price of the capacity factor is the same in the optimal subprogrammes. If the shadow prices computed by the divisions differ significantly, the allocation is revised*; new subprogrammes (with shadow prices) are calculated and subjected to the optimality test; and so on until a total optimum is attained after a number of iterations.

The second procedure starts with a *tentative shadow price fixed by the central authority*. This price is *used as an accounting price at which the divisions "buy" the capacity units*, i.e., the preference functions f_1 and f_2 are reduced by the "costs" of using the common capacity factor, evaluated at the shadow price; in this way the divisions are forced to economize on

---

1 *Exercise:* Indicate the optimal total solution in the case where the right-hand side of (8) is 12 instead of 8 (cf. Fig. 20).

their use of capacity. Optimal subprogrammes are then computed with the modified preference functions. *If the divisions still use too much of the capacity factor, or if there is now excess capacity, the accounting price is revised*, and so forth until a total optimum is attained.

We shall now proceed to show how algorithms based on these two principles can be applied to the example[1].

B. Decomposition by Direct Allocation

1. The first method[2] starts with *a tentative allocation of the common capacity*, the right-hand side of (8). Let *3* units (machine hours) be allocated to division 1, and *5* units to division 2; the divisions are informed of this allocation.

The first division now solves its subproblem,

$$f_1 = 10\, x_1 = \max$$

$$2\,x_1 \leq 8 \quad \text{or} \quad 2\,x_1 + x_3 \quad\quad = 8 \tag{6}$$

$$1\,x_1 \leq 3 \quad \text{or} \quad 1\,x_1 \quad\quad + x_6 = 3, \tag{8a}$$

where x_6 is a slack variable. The optimal simplex tableau is:

		x_0	x_1	x_3	x_6
	x_1	3	1	0	1
Ia	x_3	2	0	1	-2
	f_1	30	0	0	10

The capacity allocated to the division is seen to be fully utilized ($x_6 = 0$); hence the shadow price — representing the division's internal valuation of a marginal unit of capacity — will be positive. It is equal to the simplex coefficient of the slack variable x_6,

$$y_1 = -\partial f_1/\partial x_6 = 10.$$

Similarly, the subproblem of division 2 (with a slack variable x_7),

$$f_2 = 10\, x_2 = \max$$

$$3\,x_2 + x_4 \quad\quad = 9 \tag{7}$$

$$2\,x_2 \quad\quad + x_7 = 5, \tag{8b}$$

---

[1] The simple algorithms used in sections B–C below are based on I. Thygesen (1971), Ch. VII. 2–3 and Ch. IV.5. They have been selected here because of their graphic simplicity in explaining what decentralization is about rather than for computational efficiency.

[2] The procedure may be thought of as a simplified version of an algorithm for two-level economic planning developed by Kornai and Lipták; cf. J. Kornai and T. Lipták (1965) and J. Kornai (1967), Ch. 24 and Appendix, H.

has the solution

		x_0	x_2	x_4	x_7
Ib	x_2	2.5	1	0	0.5
	x_4	1.5	0	1	−1.5
	f_2	25	0	0	5

with the shadow price

$$y_2 = -\partial f_2/\partial x_7 = 5.$$

These shadow prices are reported by the divisions to the central authority, which will now act on the following *decision rule*: *change the allocation in favour of the division having the higher shadow price* — in this case division 1 since $y_1 > y_2$. The shadow price can obviously be interpreted as the amount by which a division can increase its profit if its allocation is increased by a marginal unit of capacity; hence, changing the allocation from (3, 5) to $(3+\Delta, 5-\Delta)$ where $\Delta > 0$, total profit $f_1 + f_2$ will increase by the positive amount

$$10 \cdot \Delta - 5 \cdot \Delta = 5 \cdot \Delta \quad (>0).$$

Let the allocation — i.e., the right-hand sides of (8a) and (8b) — be changed into 5 and 3 ($\Delta = 2$); this is reported to the divisions, which now solve the revised subproblems. The results are for division 1

		x_0	x_1	x_3	x_6
IIa	x_1	4	1	0.5	0
	x_6	1	0	−0.5	1
	f_1	40	0	5	0

and for division 2

		x_0	x_2	x_4	x_7
IIb	x_2	1.5	1	0	0.5
	x_4	4.5	0	1	−1.5
	f_2	15	0	0	5

The corresponding shadow prices — the simplex coefficients in the columns of x_6 and x_7 — are

$$y_1 = 0, \quad y_2 = 5^1,$$

i.e., $y_2 > y_1$, which indicates that the central authority in adjusting the allocation has overshot the mark. According to the decision rule, division 2

---

[1] The capacity factor is not fully utilized in division 1, hence another unit would add nothing to profit and $y_1 = 0$.

is to have its allocation cut down — though of course not to the original level.

Let us try the allocation *4* to division 1 and *4* to division 2. Then the subproblem of division 1 will have two alternate degenerate solutions:

		x_0	x_1	x_3	x_6
IIIa	x_1	4	1	0	1
	x_3	0	0	1	−2
	f_1	40	0	0	10
IVa	x_1	4	1	0.5	0
	x_6	0	0	−0.5	1
	f_1	40	0	5	0

The two solutions imply the same production programme ($x_1 = 4$, $f = 40$) and differ only with respect to the choice of zero variable in the basis. But the shadow price is $y_1 = 10$ in Tableau IIIa, whereas $y_1 = 0$ in Tableau IVa[1]. Hence the complete solution for y_1 is any convex combination of the two values, i.e., any value in the interval

$$0 \leq y_1 \leq 10.$$

For division 2 we have

		x_0	x_2	x_4	x_7
IIIb	x_2	2	1	0	0.5
	x_4	3	0	1	−1.5
	f_2	20	0	0	5

with the shadow price

$$y_2 = 5.$$

The shadow prices are now in agreement, the value of y_2 being in the interval for y_1, so we have solved the total problem. Tableaux IIIa—b give the optimal total solution

$$x_1 = 4, \quad x_2 = 2; \quad f = f_1 + f_2 = 60^2. \tag{9}$$

---

[1] The dual of a problem whose solution is degenerate has multiple optimal solutions, and *vice versa*.

The interpretation of alternate shadow prices in this case is that a unit *less* of the capacity factor would reduce profit by 10 (cf. also Tableau Ia) whereas a unit *more* would be worth nothing because it would not be used anyway (cf. Tableau IIa).

[2] *Exercise:* Check this result by solving the total problem by the simplex method without decomposing it.

The method used is readily generalized to cases with any number of divisions, variables, and restrictions. It can be described as a *primal decomposition algorithm* because the iterations represent feasible solutions to the primal problem. The procedure can be elaborated into a computer algorithm [1].

2. The solution was attained through a converging sequence of solutions to the subproblems, resulting from successive adjustments of the allocation. If the allocation is considered given up to a parameter v, where v and $8-v$ units respectively are allocated to the two divisions, *the subproblems can be solved parametrically* [2]. For *division 1* we get the following parametric solution where the optimal basis and hence the shadow price changes at the critical value $v=4$:

	x_0	x_1	x_3	x_6	
x_1	v	1	0	1	feasible
x_3	$8-2v$	0	1	-2	for $v \leq 4$
f_1	$10v$	0	0	10	
x_1	4	1	0.5	0	feasible
x_6	$v-4$	0	-0.5	1	for $v \geq 4$
f_1	40	0	5	0	

For *division 2* we find similarly:

	x_0	x_2	x_4	x_7	
x_2	$4-0.5v$	1	0	0.5	feasible
x_4	$1.5v-3$	0	1	-1.5	for $v \geq 2$, i.e.
f_2	$40-5v$	0	0	5	for $8-v \leq 6$
x_2	3	1	1/3	0	feasible
x_7	$2-v$	0	$-2/3$	1	for $v \leq 2$, i.e.
f_2	30	0	10/3	0	for $8-v \geq 6$

The corresponding shadow prices y_1 and y_2 can be shown in a diagram as functions of the parameter v (or, read from right, as functions of $8-v$, the capacity allocated to division 2), cf. Fig. 21.

---

1 Cf. I. Thygesen, *op. cit.*, p. 259.
2 Using the procedures for parametric programming described in Ch. VII, C.

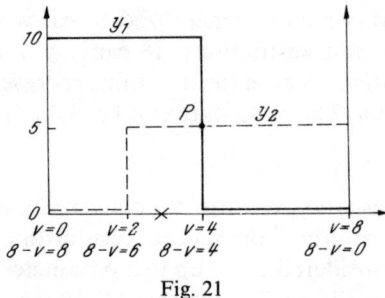

Fig. 21

To the left of $v=4$ — starting at $v=0$ — we have $y_1 > y_2$; division 1 will profit by paying the price 10 per unit for having more capacity allocated to it, while division 2 is willing to relinquish the first 2 of its 8 factor units free of charge and can afford to "sell" the next two units at a price of 5 per unit. To the right of $v=4$ the position has been reversed.

In other words, the curves may be thought of as the "demand" curve of division 1 and the "supply" curve of division 2; point P ($v=4$, $8-v=4$) represents an equilibrium where the "price" per unit is 5. The iterative procedure by which we solved the problem can therefore be interpreted as *a simulated market mechanism* by which — starting from an arbitrary allocation — a market equilibrium is attained through a series of adjustments. The algorithm simulates a trade in capacity units between the divisions.

C. Decomposition by Shadow Prices

1. In the second algorithm the allocation of the common capacity factor is managed indirectly through *a shadow price y which the divisions have to "pay" for capacity used*. According to (8) the divisions use 1 x_1 and 2 x_2 units of the capacity factor respectively, for which they are charged the price y per unit. Deducting these "costs" from the profit function, and leaving out the common restriction (8) which is no longer to be used for direct control, we have the following *"modified total problem"*:

$$f^* = (10 - 1 \cdot y) x_1 + (10 - 2 \cdot y) x_2 = \max$$

subject to the restrictions

$$2 x_1 \leq 8 \qquad (6)$$
$$3 x_2 \leq 9 \qquad (7)$$

where y is provisionally considered as a parameter. It can be proved[1] that

---
1 See Appendix, F.

the optimal solution to the original total problem is contained in the complete solution to the modified problem, namely, for the value of y which corresponds to the optimal shadow price of the common factor.

The modified total problem contains no restrictions connecting x_1 and x_2, so it can be decomposed completely into two independent subproblems

$$f_1^* = (10-y)x_1 = \max \\ 2x_1 \leq 8 \quad \text{or} \quad 2x_1 + x_3 = 8 \tag{10}$$

and

$$f_2^* = (10-2y)x_2 = \max \\ 3x_2 \leq 9 \quad \text{or} \quad 3x_2 + x_4 = 9. \tag{11}$$

Now the central authority fixes *a tentative value of y*, say, $y=4$, and informs the divisions of this accounting price. Then the divisions solve their respective subproblems (10) and (11) with $y=4$. The optimal solutions are given in Tableaux I'a and I'b:

		x_0	x_1	x_3
I'a	x_1	4	1	1/2
	f_1^*	24	0	3

		x_0	x_2	x_4
I'b	x_2	3	1	1/3
	f_2^*	6	0	2/3

The solutions are reported to the central authority. After inserting them in the common restriction (8), the central authority applies the following *decision rule: raise the shadow price if the common restriction is violated* (i.e., the solution is not feasible); *lower the price if the capacity is less than fully utilized* by the combined subprogrammes (i.e., the solution is not optimal because a slack variable and the corresponding shadow price cannot both be positive in the optimum[1]). In this case, with $y=4$, we have

$$1 \cdot x_1 + 2 \cdot x_2 = 10 > 8$$

so that solutions I'a—I'b violate (8). Hence the shadow price is raised, for example, to $y=6$.

With $y=6$ in the subproblems the divisions compute the following solutions and report them back:

		x_0	x_1	x_3
II'a	x_1	4	1	1/2
	f_1^*	16	0	2

		x_0	x_2	x_4
II'b	x_4	9	3	1
	f_2^*	0	2	0

Checking whether these solutions satisfy (8), the central authority finds that

---

[1] This follows directly from the Duality Theorem.

$$1 \cdot x_1 + 2 \cdot x_2 = 4 < 8$$

so that the slack variable of (8) is positive, which is incompatible with optimality since the shadow price was positive too. Hence the shadow price will have to be lowered, though not to the original level of 4; let it be reduced to $y = 5$, say. With this accounting price the subprogrammes become

III′a

	x_0	x_1	x_3
x_1	4	1	1/2
f_1^*	20	0	5/2

III′b

	x_0	x_2	x_4
x_2	3	1	1/3
f_2^*	0	0	0

IV′b

	x_0	x_2	x_4
x_4	9	3	1
f_2^*	0	0	0

Now the subproblem of division 2 has alternate optima, the complete solution being

$$x_2 = \theta \cdot 3 + (1 - \theta) \cdot 0 = 3\,\theta$$
$$x_4 = \theta \cdot 0 + (1 - \theta) \cdot 9 = 9 - 9\,\theta \qquad (0 \leq \theta \leq 1).$$

The left-hand side of (8) then becomes

$$1 \cdot x_1 + 2 \cdot x_2 = 4 + 6 \cdot \theta,$$

which is equal to the right-hand side ($=8$) for $\theta = 2/3$, i.e., for $x_2 = 2$ (and $x_4 = 3$). Thus it has become possible at this stage to balance the common restriction such that the capacity is fully utilized ($x_5 = 0$) and the corresponding shadow price is positive, as required of an optimal total solution. This completes the iterative procedure; the optimal overall solution is $x_1 = 4$, $x_2 = 2$, $x_4 = 3$, corresponding to (9) above[1].

This procedure, which may be described as a *dual decomposition method* because each iteration can be shown to represent a dual feasible solution, can be generalized to cover more complex cases (more than two divisions, many restrictions and variables) and can be worked out in detail as a computer algorithm[2]. A weak point of the method is the fact that it does not produce a feasible solution until termination, whereas a primal

---

1 It is no accident that one of the subproblems turned out to have alternate optima in the last iteration (Tableaux III′b–IV′b). The two subproblems have one side condition each, whereas the total problem has three side conditions so that three variables are required in the basis. Therefore, if the combined subprogrammes are to form an optimal basic solution with three variables, one of them *must* have two optimal basic solutions, i.e., a complete solution with two positive variables — in the present case, $x_2 = 3 \cdot \theta = 2$ and $x_4 = (1 - \theta) \cdot 9 = 3$.

2 Cf. I. Thygesen (1971), p. 263.

decomposition procedure can be interrupted at any step if the planner is satisfied with an approximate solution.

2. Again, the optimal solution could have been determined by *solving the subproblems parametrically*, the accounting price y being now the parameter. For *division 1* we get the parametric solution

	x_0	x_1	x_3	
x_1	4	1	1/2	optimal for $y \leq 10$
f_1^*	$40-4y$	0	$\dfrac{10-y}{2}$	
x_3	8	2	1	optimal for $y \geq 10$
f_1^*	0	$y-10$	0	

and for *division 2*

	x_0	x_2	x_4	
x_2	3	1	1/3	optimal for $y \leq 5$
f_2^*	$30-6y$	0	$\dfrac{10-2y}{3}$	
x_4	9	3	1	optimal for $y \geq 5$
f_2^*	0	$2y-10$	0	

For each division the parametric solution defines the division's demand for the capacity factor, $v_{31} = 1 \cdot x_1$ and $v_{32} = 2 \cdot x_2$, as a function of the accounting price y, as shown in Fig. 22.

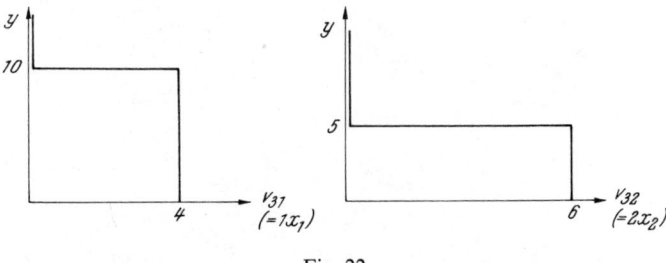

Fig. 22

Horizontal addition of the divisional demand curves gives the total "demand curve" for the capacity factor's services, $v_3 = v_3(y)$, cf. Fig. 23; the "supply curve" is the capacity limit $\bar{v}_3 = 8$. At the point of intersection, P, demand and supply balance at the price $y = 5$.

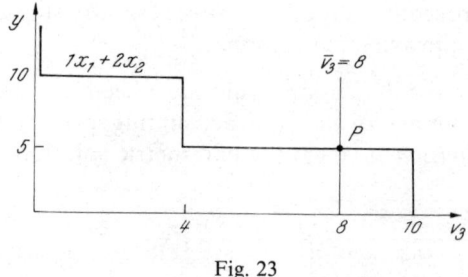

Fig. 23

The dual decomposition method can thus be thought of as a *simulated market mechanism*; the divisions "purchase" capacity from the central authority and the internal price system sees that the decentralized plans or decisions are adjusted to satisfy the overall criterion of optimality. The iterations correspond to successive market (price) adjustments which, starting from an initial disequilibrium, automatically moves the system towards equilibrium.

X. Appendix

A. Proof of the Fundamental Theorem[1]

The general linear programming problem in n variables (structural as well as slack variables) may be stated as follows:

$$A_1 x_1 + A_2 x_2 + \ldots + A_n x_n = B$$
$$x_1 \geq 0, x_2 \geq 0, \ldots, x_n \geq 0 \qquad (1)$$
$$c_1 x_1 + c_2 x_2 + \ldots + c_n x_n = f = \text{maximum}$$

where $A_1, A_2, \ldots, A_n$ and B are m-dimensional column vectors the elements of which are the coefficients of the m linear equations $(m < n)$.

Fundamental Theorem: If (1) can be solved, it will have an optimal solution in which at most m variables are positive.

We assume that we have an optimal solution of (1) in which r variables — say the r first — are positive, the remaining $n - r$ variables being zero. If $r \leq m$, the theorem is automatically true, so let us assume $r > m$. Let

$$(x_1, x_2, \ldots, x_r, x_{r+1}, \ldots, x_n) = (\lambda_1, \lambda_2, \ldots, \lambda_r, 0, \ldots, 0)$$

be this solution; this means that

$$A_1 \lambda_1 + A_2 \lambda_2 + \ldots + A_r \lambda_r = B$$
$$\lambda_1 > 0, \lambda_2 > 0, \ldots, \lambda_r > 0$$
$$c_1 \lambda_1 + c_2 \lambda_2 + \ldots + c_r \lambda_r = f_{\max}.$$

Now a well-known theorem of linear algebra states that a set of vectors is always linearly dependent if their number is greater than their dimension. It follows that there will exist a set of numbers γ_j, not all zero, such that

$$A_1 \gamma_1 + A_2 \gamma_2 + \ldots + A_r \gamma_r = 0;$$

we may assume that at least one γ_j is positive, for if not we could change the sign of all γ_j and the equation would still be satisfied. Now let $\theta = \max (\gamma_j / \lambda_j)$, which is clearly a positive number. We can then easily show that

$$A_1 \left(\lambda_1 - \frac{\gamma_1}{\theta} \right) + A_2 \left(\lambda_2 - \frac{\gamma_2}{\theta} \right) + \ldots + A_r \left(\lambda_r - \frac{\gamma_r}{\theta} \right) = B,$$

---

[1] This proof of the theorem is due to David Gale.

i.e., that the set of numbers

$$\lambda_j - \frac{\gamma_j}{\theta} \quad (j=1, 2, \ldots, r)$$

is a *solution* to the equation system of (1). The *non-negativity* of the solution follows from the assumptions

$$\frac{\gamma_1}{\lambda_j} \leq \theta, \ \lambda_j > 0, \ \text{and} \ \theta > 0.$$

Moreover, by the definition of θ, at least one of the numbers is zero, i.e., we have found a feasible solution in which *fewer than r variables are positive*. The proof is now complete if we can show that the solution is also an *optimal* one, i.e., that

$$c_1 \left(\lambda_1 - \frac{\gamma_1}{\theta} \right) + c_2 \left(\lambda_2 - \frac{\gamma_2}{\theta} \right) + \ldots + c_r \left(\lambda_r - \frac{\gamma_r}{\theta} \right)$$
$$= c_1 \lambda_1 + c_2 \lambda_2 + \ldots + c_r \lambda_r \ (= f_{\max}),$$

which will be true if we can show that

$$c_1 \gamma_1 + c_2 \gamma_2 + \ldots + c_r \gamma_r = 0.$$

Supposing that this were not so we could find a number μ such that

$$\mu(c_1 \gamma_1 + c_2 \gamma_2 + \ldots + c_r \gamma_r) = c_1(\mu \gamma_1) + c_2(\mu \gamma_2) + \ldots + c_r(\mu \gamma_r) > 0$$

or, adding $\sum_{j=1}^{r} c_j \lambda_j$ on either side of the inequality sign,

$$c_1(\lambda_1 + \mu \gamma_1) + c_2(\lambda_2 + \mu \gamma_2) + \ldots + c_r(\lambda_r + \mu \gamma_r)$$
$$> c_1 \lambda_1 + c_2 \lambda_2 + \ldots + c_r \lambda_r \ (= f_{\max}).$$

The $(\lambda_j + \mu \gamma_j)$ are easily shown to be a solution to the equation system in (1) for any value of μ, and by making $|\mu|$ sufficiently small we could make it a non-negative solution, but this would mean that the $(\lambda_j + \mu \gamma_j)$ would give a larger value of f than the λ_j, contradicting the assumption that the λ_j are an optimal solution.

We have thus proved that the number of positive variables in an optimal solution can always be reduced so long as $r > m$. Repetition of this reduction must eventually lead to a solution in which at most m variables are positive.

B. The Simplex Criterion

Consider the general linear programming problem

$$AX = B, \quad X \geq 0, \quad C^* X = f = \text{maximum} \tag{2}$$

where A is the matrix of coefficients a_{ij} (of dimension $m \times n$), X the column vector of the variables (including slack variables), and B the column vector of the b_i, whereas C^* is a row vector whose elements are the c_j. The matrix A can be partitioned into a basis A_1, consisting of m linearly independent column vectors, and the matrix of non-basic activities, A_2. (No assumption is made with regard to the distribution of the slack vectors between A_1 and A_2; slack as well as structural activities may occur in both and they are treated alike.) Let the vectors X and C be partitioned similarly. We can then rewrite (2) as

$$A_1 X_1 + A_2 X_2 = B \\ X_1 \geq 0, X_2 \geq 0 \\ C_1^* X_1 + C_2^* X_2 = f = \text{maximum}. \tag{3}$$

Solving the equation system for X_1 — which is always possible when the basic vectors are assumed to be linearly independent — we get

$$X_1 = A_1^{-1} B - A_1^{-1} A_2 X_2 \tag{4}$$

and substituting in the preference function gives f as a function of X_2:

$$f = C_1^* A_1^{-1} B + (C_2^* - C_1^* A_1^{-1} A_2) X_2. \tag{5}$$

For $X_2 = 0$ we have the basic solution corresponding to the basis A_1:

$$X_1 = A_1^{-1} B, \quad X_2 = 0, \quad f = C_1^* A_1^{-1} B. \tag{6}$$

Let us assume that it is a feasible solution, i.e., $A_1^{-1} B \geq 0$.

Now if the coefficient of X_2 in (5) — i.e., the vector of the "simplex coefficients" — is non-positive,

$$C_2^* - C_1^* A_1^{-1} A_2 \leq 0, \tag{7}$$

so that f is a non-increasing function of X_2, f will obviously assume its maximal value for $X_2 = 0$, negative values to the elements of X_2 not being admissible. In other words, (7) is a *sufficient* condition of a maximum (the "*Simplex Criterion*") and if it happens to be satisfied we have solved the problem in a single step. In a minimization problem the inequality sign in (7) must of course be reversed.

If the simplex criterion is not satisfied for the basis considered, i.e., if there is at least one positive simplex coefficient, say the coefficient of the variable x_k, we can get a better solution by giving x_k a small positive value so long as X_1 remains non-negative. This can always be done when the initial basic solution is strictly positive, $A_1^{-1} B > 0$[1]. However, if f has a finite maximum, x_k cannot be increased indefinitely; sooner or

---

[1] When this assumption is made it involves no further restriction to assume that the m column vectors in A_1 are linearly independent (as we did, to ensure that the matrix A_1 could be inverted). If the equation $A_1 X_1 = B$ can be solved, linear dependence in A_1 would imply the existence of a degenerate solution.

later one of the variables in X_1 becomes zero — inspection of (4) will show when it first happens — and this point represents a new basis in which one of the original basic vectors has been replaced by one from A_2. If the simplex criterion is satisfied for the new basis, the corresponding basic solution is optimal; if not, we go on to a third basis, and so forth. Since the value of f increases each time the basis is changed, a basis which has once been left can never reappear at a later stage of the procedure. Since, moreover, the number of bases is finite, it follows that the method will eventually lead to an optimal solution (provided such a solution exists) in a finite number of steps[1].

If we drop the assumption that the basic solution $A_1^{-1} B$ is strictly positive, i.e., if the solution is allowed to be degenerate, the above argument cannot be carried through. It is easily shown by examples that it is perfectly possible for a solution to be optimal and some of the simplex coefficients in (7) to be positive at the same time, because a non-basic variable x_k which has a positive simplex coefficient can be increased only at the expense of a basic variable which was zero in the basic solution and which will therefore become negative for positive x_k. It follows that the simplex criterion (7) is a *necessary* condition of a maximum only if the basic solution (6) is not degenerate. Moreover, the occurrence of degenerate basic solutions may prevent the convergence of the simplex procedure, but this difficulty is easily overcome by a slight modification of the simplex method, cf. Ch. V, E.

When all of the simplex coefficients are negative this means that the solution is not only optimal but unique. If there are zeroes among the simplex coefficients the solution is still optimal since (7) is satisfied, but in this case the problem will of course have more than one basic optimal solution (and any linear convex combination of these basic solutions will also be optimal).

C. The Simplex Algorithm

1. The general linear programming problem with m restrictions in n variables (including slack variables),

$$
\begin{aligned}
& a_{11} x_1 + a_{12} x_2 + \ldots + a_{1n} x_n = b_1 \\
& a_{21} x_1 + a_{22} x_2 + \ldots + a_{2n} x_n = b_2 \\
& \ldots \\
& a_{m1} x_1 + a_{m2} x_2 + \ldots + a_{mn} x_n = b_m \\
& x_1 \geq 0, \; x_2 \geq 0, \ldots, x_n \geq 0 \\
& c_1 x_1 + c_2 x_2 + \ldots + c_n x_n = f = \text{maximum},
\end{aligned}
\tag{8}
$$

---

[1] Note that for $m=1$, $n=2$ the vectors and matrices of the problem (3) above become scalars and the simplex procedure becomes identical with the one we used in solving the trivial problem (4) in Ch. II.

The Simplex Algorithm

is solved in a number of steps by successively changing the basis until we get to a basis which satisfies the simplex criterion. We will now prove the formulas by which we get from one step to the next, i.e., the *algorithm* for transforming the simplex tableau.

We choose as an initial basis a set of m variables $(x_r, ..., x_i, ...)$ where x_i is the general basic variable and x_r is the variable to be replaced in the next step (by x_k), $i \neq r$. Solving (8) for the basic variables in terms of the other $(n-m)$ variables we get

$$x_r = \bar{x}_r - x_{rj} x_j - x_{rk} x_k$$
$$x_i = \bar{x}_i - x_{ij} x_j - x_{ik} x_k \tag{9a}$$

where x_j is the general non-basic variable, $j \neq k$. (In general (for $m>2$, $n>4$) there will of course be several x_i's and several x_j terms; this being understood, (9a) says all that we need). Substituting in the linear form

$$f = c_r x_r + c_i x_i + c_j x_j + c_k x_k$$

we get
$$f = \bar{x}_0 - x_{0j} x_j - x_{0k} x_k \tag{9b}$$
where

$\bar{x}_0 = c_r \bar{x}_r + c_i \bar{x}_i$, $x_{0j} = c_r x_{rj} + c_i x_{ij} - c_j$, and $x_{0k} = c_r x_{rk} + c_i x_{ik} - c_k$.

Thus the first basic solution is

$$x_r = \bar{x}_r, \ x_i = \bar{x}_i, \ f = c_r \bar{x}_r + c_i \bar{x}_i^1.$$

We will assume that it is feasible (x_r and $x_i \geq 0$) but not optimal, that x_k is the variable now to be introduced into the basis, and that x_r is the variable to be replaced[2].

Now the second step of the simplex procedure is to solve the equations of (8) for the new basic variables (x_k, x_i) in terms of x_r and x_j. This is most easily done from (9a); x_k is found from the first equation and x_i from the second after substituting for x_k and we get

---

[1] If x_r and the x_i are slack vectors (with zero coefficients in f) they will be equal to the respective right-hand terms in (8) and f will be zero in this solution.

[2] The interpretation of the simplex coefficients $x_{0s} = z_s - c_s$ ($s = j, k$) is as follows: By (9b),

$$\frac{\partial f}{\partial x_s} = -x_{0s} = c_s - (c_r x_{rs} + c_i x_{is}),$$

which represents the net increase in f per unit of increase in x_s, starting from the basic solution in which x_s was zero. The first term, c_s, represents the "direct" effect of the increase in x_s since it is the coefficient associated with x_s in the linear preference function. The second term, $z_s = c_r x_{rs} + c_i x_{is}$, represents the aggregate negative effect on the value of f; by (9a), a unit increase in x_s will reduce x_r and x_i by the amounts x_{rs} and x_{is} respectively, and the sum of these two weighted with their associated coefficients from the preference function is z_s. The simplex criterion is concerned with the sign of the net effect $c_s - z_s$ (for convenience denoted $-x_{0s}$ above).

$$x_k = \frac{x_r}{x_{rk}} - \frac{1}{x_{rk}} x_r - \frac{x_{rj}}{x_{rk}} x_j$$

$$x_i = \left(\bar{x}_i - x_{ik} \cdot \frac{\bar{x}_r}{x_{rk}}\right) - \left(0 - x_{ik} \cdot \frac{1}{x_{rk}}\right) x_r - \left(x_{ij} - x_{ik} \cdot \frac{x_{rj}}{x_{rk}}\right) x_j.$$ (10a)

A similar substitution in (9b) gives

$$f = \left(\bar{x}_0 - x_{0k} \cdot \frac{\bar{x}_r}{x_{rk}}\right) - \left(0 - x_{0k} \cdot \frac{1}{x_{rk}}\right) x_r - \left(x_{0j} - x_{0k} \cdot \frac{x_{rj}}{x_{rk}}\right) x_j. \quad (10b)$$

The coefficients of (10a—b) indicate how the transformation from the first step to the second is carried out. The algorithm applies equally to the subsequent steps since the transformations are mere repetitions of the same formal procedure.

2. Equations (9a—b) can be rearranged as follows, with the constant terms isolated on the left-hand side:

$$\bar{x}_r = 1\, x_r + 0\, x_i + x_{rj}\, x_j + x_{rk}\, x_k$$
$$\bar{x}_i = 0\, x_r + 1\, x_i + x_{ij}\, x_j + x_{ik}\, x_k$$
$$\bar{x}_0 = 0\, x_r + 0\, x_i + x_{0j}\, x_j + x_{0k}\, x_k + f.$$

The *simplex tableau* of the first step is nothing but a table showing the coefficients of this system. The simplex tableaux corresponding to the first two steps (i.e., (9a—b) and (10a—b) respectively) are as follows:

			$c_0 (=0)$ P_0	c_r P_r	c_i P_i	c_j P_j	c_k P_k
I	c_r c_i	P_r P_i	$\bar{x}_r$ $\bar{x}_i$	$(x_{rr}=)\,1$ $(x_{ir}=)\,0$	$(x_{ri}=)\,0$ $(x_{ii}=)\,1$	x_{rj} x_{ij}	x_{rk} x_{ik}
	$z_s - c_s$		$\bar{x}_0$	0	0	x_{0j}	x_{0k}
II	c_k	P_k	$\dfrac{\bar{x}_r}{x_{rk}}$	$\dfrac{1}{x_{rk}}$	0	$\dfrac{x_{rj}}{x_{rk}}$	1
	c_i	P_i	$\bar{x}_i - x_{ik} \cdot \dfrac{\bar{x}_r}{x_{rk}}$	$0 - x_{ik} \cdot \dfrac{1}{x_{rk}}$	1	$x_{ij} - x_{ik} \cdot \dfrac{x_{rj}}{x_{rk}}$	0
	$z_s - c_s$		$\bar{x}_0 - x_{0k} \cdot \dfrac{\bar{x}_r}{x_{rk}}$	$0 - x_{0k} \cdot \dfrac{1}{x_{rk}}$	0	$x_{0j} - x_{0k} \cdot \dfrac{x_{rj}}{x_{rk}}$	0

In Tableau I, the simplex coefficients in the columns P_j and P_k are computed as

$$x_{0s} = c_r\, x_{rs} + c_i\, x_{is} - c_s = z_s - c_s \quad (s=j, k);$$

this relation also holds in the basis columns ($s=r, i$) as well as for $s=0$, if, for convenience, we write $c_0=0$ at the top of the P_0 column. Furthermore, as the reader may easily verify, it holds in each column of Tableau II, thus giving us a check on the calculations in the last line.

D. Proof of the Duality Theorem

1. A linear programming problem with *m inequalities* in *n structural* variables,

$$AX \leq B, \quad X \geq 0, \quad C^*X = f = \text{maximum}, \qquad (11\,\text{a})$$

where A is the $(m \times n)$ matrix of coefficients a_{ij}, X and B the column vectors of the x_j and b_i respectively, and C^* the row vector of the c_j, has a *dual* with n inequalities in m structural variables:

$$A^*Y \geq C, \quad Y \geq 0, \quad B^*Y = g = \text{minimum}, \qquad (12\,\text{a})$$

where A^* is the transpose of A, and Y is a column vector of variables $y_1, y_2, \ldots, y_m$. The two problems can be rewritten as

$$AX + I_m X' = B, \; X \geq 0, \; X' \geq 0, \; C^*X = f = \text{maximum} \qquad (11\,\text{b})$$

$$A^*Y - I_n Y' = C, \; Y \geq 0, \; Y' \geq 0, \; B^*Y = g = \text{minimum}, \qquad (12\,\text{b})$$

where I_m and I_n are unit matrices, and X' and Y' are column vectors of slack variables $x'_1, x'_2, \ldots, x'_m$ and $y'_1, y'_2, \ldots, y'_n$ respectively.

Just as the two problems are symmetrical in their construction, there is a close correspondence ("duality") between their respective solutions, as stated by the

Duality Theorem: In the optimal basic solutions of (11b) and (12b):

1° The values of x_j and x'_i in the basis are numerically equal to the simplex coefficients of y'_j and y_i, and the values of y_i and y'_j in the basis are numerically equal to the simplex coefficients of x'_i and x_j respectively.

2° $f_{\max} = g_{\min}$.

2. Let us first assume that the optimal basis of (11b) consists of *structural* vectors only. We can then partition the vector X into the vector of *m* basic variables, X_1, and the vector of the remaining structural variables, X_2; let A and C^* be partitioned correspondingly and (11b) can be written as

$$A_1 X_1 + A_2 X_2 + I_m X' = B$$
$$X_1 \geq 0, \; X_2 \geq 0, \; X' \geq 0$$
$$C_1^* X_1 + C_2^* X_2 = f = \text{maximum}.$$

Solving the equation system for X_1 we get

$$X_1 = A_1^{-1} B - A_1^{-1} A_2 X_2 - A_1^{-1} X' \qquad (13)$$

and substituting in f gives

$$f = C_1^* A_1^{-1} B - (C_1^* A_1^{-1} A_2 - C_2^*) X_2 - C_1^* A_1^{-1} X'. \tag{14}$$

Since we have assumed that A_1 is the optimal basis, (13)—(14) give the optimal solution for $X_2 = X' = 0$.

Now (12b) can be written, after partitioning I_n and Y',

$$\begin{aligned}
A_1^* Y - I_m Y_1' &= C_1 \\
A_2^* Y \quad - I_{n-m} Y_2' &= C_2 \\
Y \geq 0, \; Y_1' \geq 0, \; Y_2' &\geq 0 \\
B^* Y = g = \text{minimum}.
\end{aligned}$$

Part 1° of the Theorem suggests the choice of (Y, Y_2') as a basis. Solving for these variables we get

$$\begin{aligned}
Y &= (A_1^*)^{-1} C_1 + (A_1^*)^{-1} Y_1' \\
Y_2' &= [A_2^* (A_1^*)^{-1} C_1 - C_2] + A_2^* (A_1^*)^{-1} Y_1'
\end{aligned} \tag{15}$$

and substitution in g gives

$$g = B^* (A_1^*)^{-1} C_1 + B^* (A_1^*)^{-1} Y_1'. \tag{16}$$

The basic solution for Y and Y_2' obtained from (15) is non-negative since it is equal to the (transposed) simplex coefficients of X' and X_2 in (14)[1]. And it is optimal because the simplex coefficient of Y_1' is positive, being equal to the (transposed) basic solution for X_1 in (13). Furthermore, the constant terms in (14) and (16) are the transposes of each other, i.e., $f_{\max} = g_{\min}$. The proof is now complete.

3. However, we have not allowed for the possibility of slack variables appearing in the optimal basis of (11). To prove the theorem in this general case let us rewrite (11) as

$$\begin{aligned}
A_1 X_1 + A_2 X_2 &= B \\
X_1 \geq 0, \; X_2 &\geq 0 \\
C_1^* X_1 + C_2^* X_2 &= f = \text{maximum},
\end{aligned} \tag{17}$$

where X_1 corresponds to the basis A_1 whereas X_2 now stands for the remaining n variables; there may be slack variables in both. Solving for X_1 and f in terms of X_2 we have

$$X_1 = A_1^{-1} B - A_1^{-1} A_2 X_2 \tag{18}$$

$$f = C_1^* A_1^{-1} B - (C_1^* A_1^{-1} A_2 - C_2^*) X_2. \tag{19}$$

---

[1] We assume that degeneracy does not occur since otherwise the simplex criterion is not a *necessary* condition for an optimum solution.

The problem:
$$A_1^* Y - I_m Y_1' = C_1$$
$$A_2^* Y \qquad - I_n Y_2' = C_2$$
$$Y \geq 0, \ Y_1' \geq 0, \ Y_2' \geq 0 \qquad (20)$$
$$B^* Y = g = \text{minimum}$$

is the dual of (17) except that equations corresponding to the m *slack* vectors in A_1 and A_2 have been added; these equations will obviously have the form

$$1 \cdot y_i - 1 \cdot y_k' = 0 \quad (i = 1, 2, \ldots, m), \qquad (21)$$

i.e., they merely identify $y_1, y_2, \ldots, y_m$ with m from among the $(n+m)$ slack variables y_k'. Solving (20) for Y, Y_2', and g in terms of Y_1' we get

$$Y = (A_1^*)^{-1} C_1 + (A_1^*)^{-1} Y_1'$$
$$Y_2' = A_2^* (A_1^*)^{-1} C_1 - C_2 + A_2^* (A_1^*)^{-1} Y_1' \qquad (22)$$
$$g = B^* (A_1^*)^{-1} C_1 + B^* (A_1^*)^{-1} Y_1'.$$

For $Y_1' = 0$ this is a feasible and optimal solution of (20) since the value of Y_2' corresponds with the simplex coefficient of X_2 in (19) whereas the simplex coefficient of Y_1' is equal to the value of X_1, cf. (18); the non-negativity of Y follows from (21). Moreover, $f_{\max} = g_{\min}$.

We have thus paired X_1 with Y_1' and Y_2' with X_2, element for element. By (21), we can replace those m elements in Y_1' and Y_2' that correspond to the m slack variables in X_1 and X_2, by the elements of Y. If the kth variable in X_1 or X_2 is a slack variable x_1', the corresponding variable in Y_1' or Y_2' is y_k' which, by (21), is equal to y_i. In other words, when the equations (21), which do not really belong to the dual problem, have been removed from the problem together with their slack variables y_k' after having served their purpose, the solution of the dual with which we are left corresponds to the solution of (17) as stated by the duality theorem.

The diagrams of Fig. 24 illustrate how this procedure works when applied to the problem

$$x_1 + 2x_2 + x_1' = 2$$
$$4x_1 + x_2 \qquad + x_2' = 6$$
$$x_1 \geq 0, \ x_2 \geq 0, \ x_1' \geq 0, \ x_2' \geq 0 \qquad (23)$$
$$5x_1 + x_2 = f = \text{maximum}$$

and its dual

$$y_1 + 4y_2 - y_1' = 5$$
$$2y_1 + y_2 \qquad - y_2' = 1$$
$$y_1 \geq 0, \ y_2 \geq 0, \ y_1' \geq 0, \ y_2' \geq 0 \qquad (24)$$
$$2y_1 + 6y_2 = g = \text{minimum},$$

the optimal solutions of which are

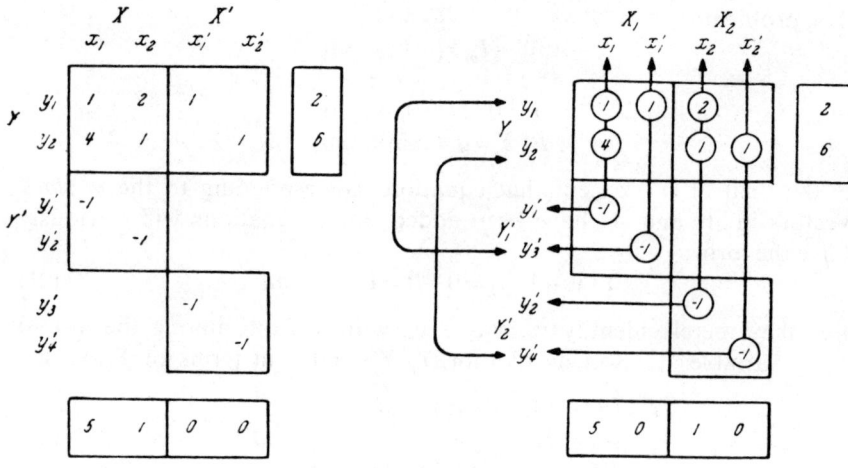

Fig. 24

$$x_1 = \frac{3}{2},\ x_1' = \frac{1}{2},\ f = \frac{15}{2}\ \text{and}\ y_2 = \frac{5}{4},\ y_2' = \frac{1}{4},\ g = \frac{15}{2}.$$

The diagram to the right illustrates (23) and the enlarged version of (24) after the column vectors of (23) have been rearranged with the basic vectors first.

E. Gomory's Algorithms for Integer Programming[1]

1. In general terms, Gomory's algorithm for *pure integer programming* can be described as follows.

Let x_i be a basic variable having a fractional value in the optimal non-integral solution. Expressed in terms of the non-basic variables x_j we then have

$$x_i = \bar{x}_i - \sum_j x_{ij} x_j$$

or, corresponding to the ith row in the simplex tableau,

$$\bar{x}_i = 1\, x_i + \sum_j x_{ij} x_j. \tag{25}$$

Now split $\bar{x}_i$ into the largest lower integer (≥ 0 because $\bar{x}_i > 0$ by assumption) and a non-negative proper fraction,

$$\bar{x}_i = N_i + f_i \quad (0 < f_i < 1),$$

and split the coefficients x_{ij} similarly:

---

[1] Cf. S. Vajda (1962), Ch. XXIV.

$$x_{ij}=N_{ij}+f_{ij} \quad (0 \le f_{ij} < 1)$$

where N_{ij} may be positive, zero, or negative (e.g., $3.7 = 3 + 0.7$; $1/11 = = 0 + 1/11$: $-0.2 = -1 + 0.8$). Then (25) can be written

$$N_i + f_i = 1 \; x_i + \sum_j (N_{ij} + f_{ij}) \, x_j. \tag{26}$$

Now define an auxiliary variable x_s by

$$x_s = -f_i + \sum_j f_{ij} \, x_j; \tag{27}$$

introduce this equation as an additional row into the simplex tableau,

$$-f_i = 1 \, x_s - \sum_j f_{ij} \, x_j, \tag{28}$$

and introduce an additional column for x_s with the coefficient 1 in the row of x_s and all other coefficients $= 0$. Thus the original basic solution with $x_s = -f_i$ added to it constitutes a basic solution in the enlarged tableau. x_s is required to be not only ≥ 0 but integral like the other variables.

The simplex coefficients in the enlarged tableau have the proper sign since they are not affected by the changes we have made, but the basic solution is negative in x_s since $x_s = -f_i < 0$. Therefore, the original solution (i.e., the optimal non-integral solution joined by $x_s = -f_i$) is not feasible in the enlarged tableau; the new restriction is a cut which excludes a subset of the original feasible region, including the optimal non-integral solution.

On the other hand, the cut does not exclude any integral solution which was feasible in the original problem. Since $0 < f_i < 1$, $f_{ij} \ge 0$, and $x_j \ge 0$ for all j, we have

$$\sum_j f_{ij} \, x_j \ge 0$$

$$-f_i > -1,$$

from which we get by addition

$$x_s > -1.$$

It follows from (26)–(27) that

$$x_s = N_i - 1 \, x_i - \sum_j N_{ij} \, x_j$$

so that integral values of x_i and the x_j imply that x_s too is an integer, $x_s > -1$ and integral means that x_s can take the values $0, 1, 2, \ldots$ only. Hence any integral solution which is feasible in the original problem will also be feasible in the enlarged integer programming problem where an additional restriction and an additional integral variable x_s has been introduced.

The basic solution corresponding to the enlarged tableau will be dual feasible. By the dual simplex method we proceed to the next basis in which x_s has been replaced by one of the original variables. If the solution is now non-negative and integral, the problem has been solved (the dual simplex method automatically preserves the proper sign of the simplex coefficients); if not, the procedure is repeated, introducing further auxiliary variables. The method will in general converge.

2. Gomory's algorithm for *mixed integer programming* proceeds in the following manner.

In the tableau corresponding to the optimal non-integral solution, select the row of a basic variable x_i which is fractional in the solution but which is required to be integral:

$$\bar{x}_i = 1\,x_i + \sum_j x_{ij}\,x_j, \tag{29}$$

and split $\bar{x}_i$:

$$\bar{x}_i = N_i + f_i \quad (N_i \geq 0 \text{ and integer}, \ 0 < f_i < 1).$$

Instead of splitting the x_{ij} as we did in the pure case, we arrange them according to sign, writing (29) in the form

$$N_i + f_i = 1\,x_i + \sum_h x_{ih}\,x_h + \sum_k x_{ik}\,x_k \tag{30}$$

where

$$x_{ih} \geq 0, \ x_h \geq 0 \text{ and } x_{ik} < 0, \ x_k \geq 0. \tag{31}$$

Add a new row defining an auxiliary variable x_s,

$$-f_i = 1\,x_s - \sum_h x_{ih}\,x_h + \sum_k (g_i\,x_{ik})\,x_k \tag{32}$$

where $g_i = f_i/(1-f_i) > 0$, and a new column which makes x_s a basic variable (i.e., with the coefficient 1 in the x_s row and otherwise zeroes). x_s is required to be ≥ 0.

The basic solution corresponding to this enlarged tableau — i.e., the optimal non-integral solution with $x_s = -f_i$ added to it — is negative in x_s since $f_i > 0$. Hence (32) cuts off the original solution. On the other hand, (32) respects any solution to (1) that is integral in x_i, for any such solution implies $x_s \geq 0$. To prove this, rewrite (30)—(31):

$$N_i - x_i + f_i = \sum_h + \sum_k, \text{ where } \sum_h \geq 0, \ \sum_k \leq 0, \tag{33}$$

and consider a non-negative solution which satisfies (33) and the original side conditions and which is integral in x_i. If the left-hand side of (33) is ≥ 0, it can have the values $f_i, f_i+1, f_i+2, \ldots$ only because $N_i - x_i$ is an integer and $0 < f_i < 1$; therefore

$$f_i \leq \sum_h + \sum_k \leq \sum_h \quad (\text{since } \sum_k \leq 0)$$

so that

$$f_i \le \sum_h \le \sum_h -g_i \sum_k \quad \text{(since } g_i > 0, \sum_k \le 0\text{)}.$$

On the other hand, if the left-hand side of (33) is <0, it can only assume the values $f_i - 1, f_i - 2, f_i - 3, \ldots$; hence we have

$$\sum_k \le \sum_h + \sum_k \le f_i - 1 \quad \text{(since } \sum_h \ge 0\text{)}$$

from which (recalling that $g_i = f_i/(1 - f_i)$) we get

$$f_i \le -g_i \sum_k \le \sum_h -g_i \sum_k \quad \text{(since } \sum_h \ge 0\text{)}.$$

Thus we have in either case

$$f_i \le \sum_h x_{ih} x_h - g_i \sum_k x_{ik} x_k,$$

which with (32) is seen to imply $x_s \ge 0$, Q.E.D.

Applying the dual simplex method we can now proceed to the next simplex tableau. If the solution is now integral in the variables for which this is required, the problem is solved; if not, repeat the procedure, introducing another auxiliary variable, and so forth. The algorithm will in general converge, but it is not very efficient; even with only two integral variables it may take many iterations to attain the optimal solution.

F. A Decomposition Theorem

If one or more of the optimal shadow prices (values of the dual variables, y_i) is known in a linear programming problem, this information can be used to remove the corresponding side condition(s) from the primal problem, thus reducing the dimension of the problem.

Let the primal problem be

$$f = \sum_{j=1}^n c_j x_j = \max$$

$$\sum_{j=1}^n a_{ij} x_j \le b_i \qquad (i = 1, 2, \ldots, m) \tag{34}$$

$$x_j \ge 0 \qquad (j = 1, 2, \ldots, n),$$

which has the dual

$$g = \sum_{i=1}^m b_i y_i = \min$$

$$\sum_{i=1}^m a_{ij} y_i \ge c_j \qquad (j = 1, 2, \ldots, n) \tag{35}$$

$$y_i \geq 0 \qquad (i=1, 2, \ldots, m).$$

Now select one of the dual variables — the first, say — and consider it provisionally as a parameter, $\bar{y}_1$ (≥ 0). Then form a *"modified" primal problem*,

$$f^* = \sum_{j=1}^{n} (c_j - a_{1j} \bar{y}_1) x_j = \max$$

$$\sum_{j=1}^{n} a_{ij} x_j \geq b_i \qquad (i=2, 3, \ldots, m) \qquad (34^*)$$

$$x_j \geq 0 \qquad (j=1, 2, \ldots, n),$$

subtracting from each coefficient c_j in the preference function the corresponding coefficient in the first side condition multiplied by $\bar{y}_1$. This parametric primal problem has the dual (35*):

$$g^* = \sum_{i=2}^{m} b_i y_i = \min$$

$$\sum_{i=2}^{m} a_{ij} y_i \geq c_j - a_{1j} \bar{y}_1 \qquad (j=1, 2, \ldots, n) \qquad (35^*)$$

$$y_i \geq 0 \qquad (i=2, 3, \ldots, m).$$

For any *given* value of $\bar{y}_1$ we can add the constant $+b_1 \bar{y}_1$ to the preference function g^* without changing the solution, and write (35*) in a form similar to (35). The optimal solution to the parametric problem (35*) will depend on the parameter $\bar{y}_1$. If $\bar{y}_1$ is given the value corresponding to the optimal solution of (35),

$$\bar{y}_1 = y_1^0,$$

that is, *the "correct" shadow price*, the solution to (35*) will obviously be the same as the solution to (35) in the other variables too,

$$y_i = y_i^0 \qquad (i=2, 3, \ldots, m).$$

In other words, the solution to the dual (35) of the original problem is contained as a special case in the solution to the dual (35*) of the modified problem. Since (34) is equivalent to (35), and (34*) to (35*), in the sense of the duality theorem — the optimal tableaux of the primal and the dual contain the same figures — it follows that *the optimal solution to the original primal problem* (34) *is contained in the optimal solution to the parametric primal* (34*), viz., for $\bar{y}_1 = y_1^0$.

This theorem can be made use of in solving (34) if the correct shadow price y_1^0 is known; all we have to do is set $\bar{y}_1 = y_1^0$ and solve the modified problem (34*), which has one side condition less. Clearly any number of side conditions can be removed from the original problem in this manner, a fact which can be exploited in a dual decomposition procedure where the subproblems are made independent by removing the common restrictions.

References

Ackoff, R. L. (ed.), Progress in Operations Research, Vol. I. New York: Wiley. 1961.
Baumol, W. J., and T. Fabian, Decomposition, Pricing for Decentralization and External Economies. Mgmt. Science *11* (1), (1964).
Beale, E. M. L., Cycling in the Dual Simplex Algorithm, Nav. Res. Logistics Quart. *2* (4), (1955).
Beale, E. M. L., Mathematical Programming in Practice. London: Pitman. 1968.
Bierman, H., and S. Smidt, The Capital Budgeting Decision, 2nd ed. New York: Macmillan. 1966.
Bordin, A. (ed.), La programmazione lineare nell'industria. Torino: Edizioni della Unione Industriale di Torino. 1954.
Charnes, A., and W. W. Cooper, Generalizations of the Warehousing Model. Operat. Res. Quart. *6* (4), (1955).
Charnes, A., and W. W. Cooper, The Stepping Stone Method of Explaining Linear Programming Calculations in Transportation Problems. Mgmt. Science *1* (1), (1954).
Charnes, A., W. W. Cooper, D. Farr, and Staff, Linear Programming and Profit Preference Scheduling for a Manufacturing Firm. J. Operat. Res. Soc. America *1* (3), (1953).
Charnes, A., W. W. Cooper, and A. Mellon, Blending Aviation Gasolines — A Study in Programming Interdependent Activities in an Integrated Oil Company. Econometrica *20* (2), (1952).
Conway, R. W., W. L. Maxwell, and L. W. Miller, Theory of Scheduling. Reading: Addison-Wesley. 1967.
Danø, S., Industrial Production Models. Wien-New York: Springer. 1966.
Danø, S., and E. L. Jensen, Production and Inventory Planning in a Fluctuating Market. Metrika *1* (1), (1958).
Dantzig, G. B., and P. Wolfe, The Decomposition Principle for Linear Programs. Econometrica *29* (4), (1961).
Dantzig, G. B., Linear Programming and Extensions. Princeton: University Press. 1963.
Dorfman, R., Mathematical, or "Linear", Programming: A Nonmathematical Exposition. Amer. Econ. Rev. *43* (5), (1953).
Geoffrion, A. M., and R. E. Marsten, Integer Programming Algorithms: A Framework and State-of-the-Art Survey. Mgmt. Science *18* (9), (1972).
Heady, E. O., and W. Candler, Linear Programming Methods. Ames: Iowa State College Press. 1958.

Henderson, A., and R. Schlaifer, Mathematical Programing: Better Information for Better Decision Making. Harv. Busin. Rev. *32* (3), (1954).

Katzman, I., Solving Feed Problems through Linear Programming. J. Farm Econ. *38* (2), (1956).

Koopmans, T. C. (ed.), Activity Analysis of Production and Allocation (Cowles Commission Monograph No. 13). New York: Wiley. 1951.

Kornai, J., Mathematical Planning of Structural Decisions. Amsterdam: North-Holland. 1967.

Kornai, J., and T. Lipták, Two-Level Planning. Econometrica *33* (1), (1965).

Nielsen, N., Simplexmethoden. Mercantilia *1956* (2).

Perry, J. H. (ed.), Chemical Engineers' Handbook, 2nd ed. New York: McGraw-Hill. 1941.

Reinfeld, N. V., and W. R. Vogel, Mathematical Programming. Englewood Cliffs: Prentice-Hall. 1958.

Sommer, H. H., The Theory and Practice of Ice Cream Making, 4th ed. Madison: Published by the Author. 1944.

Stigler, G. J., The Cost of Subsistence. J. Farm Econ *27* (2), (1945).

Symonds, G. H., Linear Programming: The Solution of Refinery Problems. New York: Published by Esso Standard Oil Company, New York. 1955.

Thygesen, I., Investeringsplanlægning. Copenhagen: Polyteknisk Forlag. 1971.

Vajda, S., Mathematical Programming. Reading: Addison-Wesley. 1961.

Vajda, S., Readings in Mathematical Programming, 2nd ed. London: Pitman. 1962.

Index

activity 6
activity level 6
additivity 24, 38 n.
all-integer programming 121
alternate optima 8 ff., 14, 84 ff., 96 n., 103, 117, 119, 141, 146, 150, 156
artificial variable 81 ff., 93, 109, 141
auxiliary variable 125 ff., 163 ff.

balanced transportation problem 62
basic solution 10 ff.
basic variable 11
basis 11 ff., 75 ff., 155
beer blending 3 f., 42 ff., 85 ff.
binary variable 134
blending problems 3 f., 15 ff., 27 ff., 136 f.
bottleneck 33 ff., 112 n.
branch-and-bound technique 129 ff., 138

capacity 2, 35, 47 ff., 51 ff., 57, 112, 118, 137, 139 ff.
capital budgeting 66
cash flow 66
cheese, blending of 28 f.
chemical "blending" processes 32 ff.
chemical reactions 29
coefficient variations 112 ff.
combinatorial problems 134 ff.
common restriction 140, 167
common-sense methods 1 f., 20 ff., 62 n.
complementarity 22
complementary investments 69 ff.
computational short cuts 80 f., 87 f., 106
computers 68 n., 82 n., 84 n., 87, 112 n., 138
convex combination 7, 85, 146, 156
convexity 8 n.
corner 9, 77

corner solution 8 f., 14, 124
cost minimization 2 ff.
critical parameter values 111, 112, 114 ff.
cut 123 ff.
cutting plane 123 ff.
cycling 91

decentralization 140 ff.
decomposition 139 ff.
degeneracy 10, 14, 22, 41 n., 46 n., 62 n., 84 n., 88 ff., 94, 97, 118, 146, 155 n., 156, 160 n.
dependent investments 67 ff.
"detours" in computation 87
diet problem 25, 27 ff., 42, 82, 106 ff., 116, 118, 120
discrete linear programming 121
disposal activity 6
dual decomposition 150
dual feasible solution 110, 117
dual problem 24, 104 ff., 118 n., 159 ff., 165 ff.
dual simplex method 109 ff., 117 ff., 126, 141
Duality Theorem 24, 31 n., 104 ff., 149, 159 ff., 165 ff.
dummy destination 63, 65, 102 f.
dummy origin 65

effective limit 13, 35, 76 ff., 85, 88 ff., 114
extreme point 9, 77

factor of production 49
factor substitution 18, 49
feasible solution 2 f., 6
feed-mix problems 29
fertilizer, blending of 29
fictitious cost 98 ff.

Index

final product 33
fixed charges 49, 108
fixed factor of production 49, 108
fixed-cost problem 135 ff.
flow time 53 ff.
food industry, blending problems in 15 ff., 28 f.
freight costs 60 ff.
Fundamental Theorem 10, 12, 19, 62, 153 f.

Gantt chart 53 ff.
gasoline blending 30 ff., 39 ff., 136 f.
Gomory's algorithms for integer programming 124 ff., 128 f., 162 ff.

ice cream, blending of 15 ff., 107 n., 113 ff.
imputed prices 24, 106 ff.
inequalities, linear 5, 8, 25
initial basis 12, 75 ff., 81 f., 87 f., 93 ff., 101 f.
input 2, 18, 27 f., 31
integer programming 68 n., 121 ff.
integral solution 121 ff.
intermediate product 30 n., 33
inventory planning 55 ff., 102 f.
investment planning 66 ff., 135
isoquant 18
iterations, number of 87, 101 f., 106

job 52 ff.
Johnson's algorithm 54 n.
joint production 18, 59

limitational factor of production 18
linear dependence 153 ff.

machine allocation 49 ff.
machine assignment problem 51
machine capacity 47 ff.
machine substitution 49 ff.
marginal cost 107, 118
material balance 41
maximum feasible solution 6
metal alloys 30
minimization problems 5, 81 ff.

mixed integer programming 121, 127 ff., 136 f., 164 f.
mixing 3 f., 15 ff., 27 ff.
Mixture Law 38 n.
modified linear programming problem 148, 166
modified transportation method 98 ff.
multiple shift operation 49, 55 ff.
multiple solutions, see alternate optima
mutually exclusive investments 71 f., 135

necessary conditions for maximum 14, 90 f., 113, 156, 160 n.
net present value 66
non-negativity requirements 2 f., 5, 18
northwest corner rule 93 ff.

objective function 6
oil refineries, blending problems in 30 ff., 39 ff., 136 f.
operations research 2
optimal programme 6 n.
optimal solution 6
optimization, economic 2
output 2, 18, 27 f., 31
overtime work 49

parametric programming 14, 111, 115 ff., 147 f., 151 f., 166
perfect substitutes 41
perturbation 92 n.
preference function 6
prerequisite investments 69 ff., 135
price discount 26
price leadership 36
primal decomposition 147
primal problem 106, 109, 165
process 6, 49
processing time 52 ff.
production model 18, 27
production planning 1, 51 ff.
production process 27, 49
production, theory of 1 f.
profit maximization 2, 31 n.
programming matrix 6 n., 18
pure integer programming 121, 123 ff., 129 ff., 162 ff.

quality balance 41
quality specification 37 ff.

ranging 112 n.
raw materials, allocation of scarce 30 ff.
raw materials, blending of 28
raw materials, grades of 41 f.
refinery problems 30 ff., 39 ff., 136 f.
requirements vector 18
rounding 70 ff., 121 ff., 134

sales potential 35 ff.
scheduling 51 ff.
seasonal fluctuations 25, 56
second-best solution 45, 84 ff.
sensitivity analysis 112 ff.
sequence of operations 51 ff.
setup costs 30 n., 136 f.
setup time 51, 85, 137
shadow price 106 ff., 118, 142 ff., 166
side condition 2, 5
simplex algorithm 77 ff., 156 ff.
simplex coefficients 12 ff., 75 ff., 95 ff., 105 ff., 113 ff., 155 f.
simplex coefficients, interpretation of 80 f., 95, 104 ff., 157 n.
simplex criterion 14, 20, 75 ff., 90 f., 113 ff., 154 ff.
simplex method 12 ff., 19, 75 ff., 154 ff.
simplex tableau 77 ff., 95 ff., 158 f.
simulated market mechanism 148, 152
slack activity 6, 155

slack variable 6, 8, 39 n., 81, 140 f., 155, 160 ff.
stepping stone method 97 n.
steps, number of 87 n., 101 f., 106
stochastic programming 112
storage capacity 59
storage cost 56, 103
storage time 59
structural activity 6, 155
structural variable 6, 155
substitutable inputs 22
substitutable investments 71 ff.
sufficient conditions for maximum 14, 91, 113, 155

technical coefficient 49, 112
"tie" 91 f.
transportation method 93 ff.
transportation problem 60 ff., 92 ff.
tree 133 n.

unbalanced transportation problem 62 ff.

vitamin pills 29

warehousing problem 59 f.
waste product 31

zero investment 66 f., 135
0−1 programming 68 ff., 121, 134 ff.
0−1 variables 68, 134 ff.

Composed by Austro-Filmsatz, Richard Gerin, A-1020 Wien
Printed by Paul Gerin, A-1021 Wien